BROOKFIELDS SCHOOL
SAGE ROAD
TILEHURST RG3 6SW

Book two

Steps in Geography

R Bateman
and
F Martin

Stanley Thornes (Publishers) Ltd

Acknowledgements

Thanks to Russein Adan Isaak, Douglas Arnold and Matthew Groves for their help.

Acknowledgements are due to the following for permission to reproduce photographs used in this book.

Robert Harding Picture Library, cover

Hussein Adan Isaac p 15; J Allan Cash pp 4, 13, 52, 60, 79, 80; Australian Information Service p 35; Bangladesh High Comission p 27; Birdseye Foods p 28; British Antarctic Survey p 6; Colorado Travel Section p 54; Food and Agricultural Organization of the United Nations pp 10, 16, 19, 70; Robert Harding Picture Library pp 4, 56, 68; Alan Hutchinson p 44; Iraq Petroleum p 11; Novosti Press Agency p 39; Ohio Department of National Resources p 90; Oxfam p 66; Petroleos Mexicanos p 30; Port of Bristol Authority p 20; Port of Vancouver p 46; Space Frontiers p 42; Torbay Borough Council p 90; Welsh Development Agency p 75; West Air Photography p 58.

Appeal poster courtesy of the Disaster Emergency Committee: p.65;
Newspaper article courtesy of The Guardian: p.66

© R. Bateman and F. Martin 1981, 1987

All rights reserved. No part of this publication may be reproduced or transmitted in any form or by any means, electronic or mechanical, including photocopy, recording, or any information storage and retrieval system, without permission in writing from the publisher or under licence from the Copyright Licensing Agency Limited. Further details of such licences (for reprographic reproduction) may be obtained from the Copyright Licensing Agency Limited, of 33–4 Alfred Place, London WC1E 7DP.

Special note
You are permitted to take copies of the 'Recap' section in this book so that pupils may have individual copies for homework. Such copies should be used solely by the institution concerned and should not be loaned to or used by outside parties. All material in the book remains copyright.

Originally published in 1981 by Hutchinson Education
Reprinted 1981, 1982, 1983 (twice), 1985
Second edition 1987

Reprinted in 1989 by
Stanley Thornes (Publishers) Ltd
Old Station Drive
Leckhampton
CHELTENHAM GL53 0DN

British Library Cataloguing in Publication Data

Bateman, R.
　Steps in geography.—2nd ed.
　1. Geography—Text-books—1945-
　I. Title　　II. Martin, F. (Frederick) 1948–
　910　　G128

ISBN 0 7487 0260 1

Printed and bound by Scotprint, Musselburgh, Scotland

Note to teachers
This series has been written to introduce students to some of the basic vocabulary, ideas and skills in geography in a way which makes the teacher's classroom task straightforward. It is not however intended that the textbooks should be the student's sole resource or that the activities included should represent the total learning strategies of a course in geography. It is expected that, in many cases, the materials will serve simply as a starting point from which to develop a variety of more active learning techniques, thus ensuring a more comprehensive and balanced set of learning experiences.

Contents

People and land/Introduction

Every place has its own special character. This character is formed by the land and by people. There are millions of separate places on earth.

Do you live somewhere like Milford Sound in New Zealand (photograph **A**)? This is a beautiful place, but it is empty of people. Living in a place like this can be a struggle to **survive**. The lives of the few people who live here are dominated by the natural things around them. The great mountains, deserts and ice caps are other places where the land rules people's lives.

Or is your home in a place like Hong Kong (photograph **B**)? There is nothing natural in this scene. Instead of woodland and grassland there are shops, factories, offices and roads. A landscape like this is dominated by people.

The chances are that the place where you live is quite different from either of the photographs. This is because of the rich variety of **scenery** and ways of life on our planet. Some people live as their ancestors did, thousands of years ago. Others live in ways that their grandparents could not have imagined.

People have changed some places for the better, as the Dutch have changed their coastline. On the other hand, we have made some places very unpleasant to live in. They are said to be **polluted**.

The units in this book show you a few of the many places and ways of life on our planet.

A

1 Read the opposite page. Write fifty or more words to describe the place you live in, as if you are describing it to a space traveller. Describe the scenery, the people, and something of their way of life. Sketch **C** may help.

C

> Where I live is not really high up, but the land is not flat. It is hilly, with a river running through the middle. The people here are English. Most of them are white, but some are brown and some black. Most people work in the local town, in factories, offices and shops.

2 Write a sentence to answer each of these questions.
   a What is a place?
   b What does 'dominate' mean?
   c In what sort of places does nature dominate?
   d In what sort of places do people dominate?
   e What does 'survive' mean?

3 Complete the crossword **D** with some of the most famous place names in the world.

D

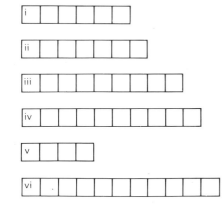

Clues
   i The capital city of China (6).
   ii The river between South Africa and Zimbabwe (7).
   iii The highest mountain in Argentina (9).
   iv The state which contains Los Angeles and San Francisco (10).
   v This huge lake is in South Australia (4).
   vi France, Germany, Italy and Austria are this country's neighbours (11).

4 Look at sketches **E** and **F**. They show two places in very different parts of the world. The words at the top describe the scene. See if you can use your imagination to draw sketches of the places that are described below:
Place X: mountains, frightening, cliffs, lake, fishermen.
Place Y: coast, very hot, sand, sea, holidaymakers.

E Lowland, peaceful, farmers, fields, river

F Flat plain, endless, thorn trees, grass, nomads, camels

5 Have a look through this book. Try to describe one of the places shown in a photograph or sketch.

Summary
The character of a place is formed by the land and by people. Natural things may dominate, so that people are hardly noticed, or the landscape may be almost all man-made. Every place on earth is different from every other one, and yet most places have something in common with others.

Unit 1.2
The frozen lands

Most young people like sliding on ice and playing in the snow. In Britain they cannot do this very often. The temperature outside must be at freezing point (0 degrees Centigrade) or below.

Near the **North** and **South Poles (polar regions)**, it is nearly always below freezing point. Sometimes the temperature drops to 40 degrees below zero (– 40°C). This is so cold that people can accidentally snap off a finger without pain. Unless people have shelter or warm clothes, they quickly die from **exposure** in the freezing wind.

Ice and snow which covers an area all year round is called an **ice cap**. Places where the ice melts and plants grow in the short summer are called **tundra** areas.

A The summer and winter tundra landscape

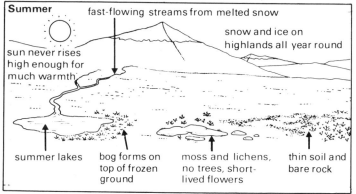

Summer

sun never rises high enough for much warmth

fast-flowing streams from melted snow

snow and ice on highlands all year round

summer lakes

bog forms on top of frozen ground

moss and lichens, no trees, short-lived flowers

thin soil and bare rock

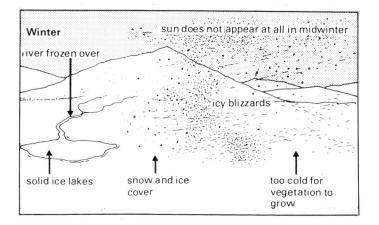

Winter

river frozen over

sun does not appear at all in midwinter

icy blizzards

solid ice lakes

snow and ice cover

too cold for vegetation to grow

Few people live in tundra areas. In the past, small groups have survived by hunting and fishing. It is too cold to grow crops and the soil is too hard. Today the hunters use guns instead of spears. They travel by motor sleds instead of using reindeer or husky dogs. Some do not work at all. Their government gives them money to buy food and clothes. Many grow tired of the hardships and move away.

B Traditional people in the northern tundra

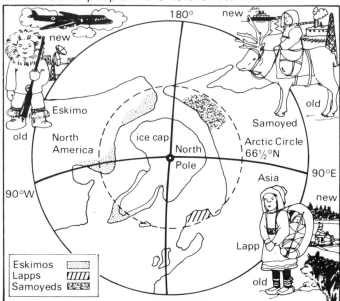

new

180°

new

Eskimo

old

North America

ice cap

North Pole

Samoyed

Arctic Circle 66½°N

90°W

90°E

Asia

new

Lapp

old

Eskimos
Lapps
Samoyeds

Ice cap areas are not completely useless. The Antarctic continent is covered by ice that is one mile thick. However, in the ground beneath there are oil, gas, coal and iron ore. There may also be gold, copper and uranium.

The ice itself locks up 90% of the earth's fresh water. There are plans to tow giant icebergs to places where farmers need water for their crops and animals.

The Southern Ocean teams with fish. A small shrimp called krill could become an important new food.

Scientists from many countries come to study the polar areas. Some want to learn about the weather. The rocks, plants and wildlife are also being studied. Before long the scientific teams may be joined by fishermen, miners and other workers.

C

1 Copy the two sketches (**A**) into your book.

2 Write out and complete each of these sentences. Add more sentences if you need to.
a Three groups of people who live in tundra areas are ....
b Modern hunters in the tundra use ....
c People move away from tundra areas because ....
d The Antarctic will become important because there are ....
e People who come to live in polar areas have jobs as ....

3 Graph **D** shows typical temperature figures for Britain and in the tundra.
a Copy the chart below and fill in the spaces with the right temperature figure.

	Britain	tundra
Warmest month temperature		
Coldest month temperature		
Number of months which are below freezing point		

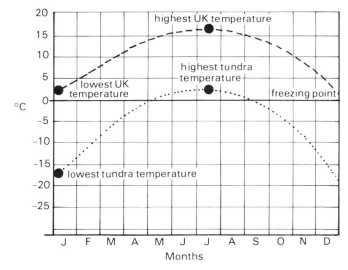

D Average UK   tundra temperatures

b Pair up the following words with the temperatures in the chart below:
very hot, cold, mild, warm, bitterly cold, hot, very cold.

degrees C	description
20 to 25	....
15 to 19	....
10 to 14	....
5 to 9	....
0 to 4	....
−5 to −1	....
−10 to −40	....

4 For a scientist in the Antarctic, life is like that of an astronaut (spaceman) in many ways.
a Study diagram **E**, then list five ways in which an Antarctic scientist lives like an astronaut.
b Design a building which you think would be suitable for the Antarctic. Draw and label in its main features.

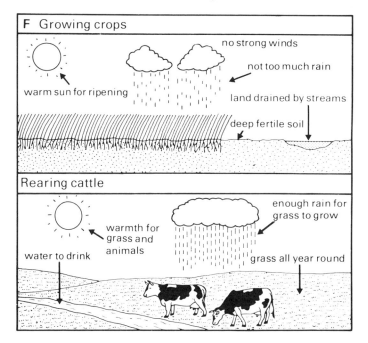

E A house for an Antarctic scientist

Advanced work

5 Diagram **F** shows the kind of conditions which crops and animals need to grow.

F Growing crops

Rearing cattle

Write a paragraph to explain why growing crops and rearing animals is almost impossible in tundra areas. Use information from sketches **A** and the temperature graph in your answer.

Summary
It is very difficult for man to live in the frozen polar lands. These areas may become more important in the future, so more people may live there.

Unit 1.3
Living on mountains

Living in mountain country has always been difficult. It is impossible to grow crops on steep rocky slopes. It is difficult to rear healthy animals in cold, wet, windy weather. Few people live in such rugged places.

Most highland people live in villages and farms in sheltered valleys. There are few towns. Towns need more space to grow in and they must be easy to get to.

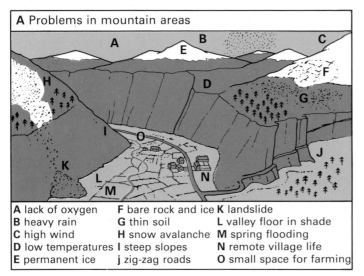

A Problems in mountain areas

A lack of oxygen	F bare rock and ice	K landslide
B heavy rain	G thin soil	L valley floor in shade
C high wind	H snow avalanche	M spring flooding
D low temperatures	I steep slopes	N remote village life
E permanent ice	j zig-zag roads	O small space for farming

The highest mountain peaks can rise above the clouds. Some peaks are so high and cold that the snow and ice never melt. When a climber crosses onto this permanent snow cover, he is said to have 'crossed the snow line'. Snow and rock sometimes crash down in an avalanche.

Map **B** shows the world's highest mountains. A line of mountains is called a **mountain range**. Where mountain ranges link up, the whole lot is called a **mountain chain**. The Andes in South America are a mountain chain.

Mountains do not last for ever. Rain and ice break off small pieces of rock. Streams carry the broken pieces away. This happens slowly but surely over millions of years.

The streams join up to become rivers. The rivers cut deep valleys into the mountain sides. A very deep and narrow valley is called a ravine. A small river which flows into a larger river is called a **tributary**.

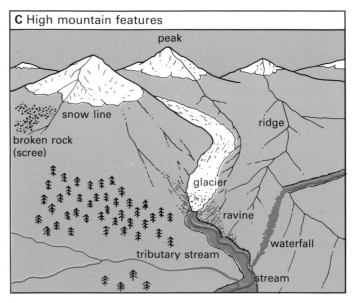

C High mountain features

In very cold and high places, ice masses called **glaciers** creep slowly down the valleys. As they move, they freeze onto the sides and drag pieces of rock with them. The ice melts as the glacier reaches lower and warmer ground.

All the pieces which are broken off and carried away end up as mud in the seas, or spread over lowland valleys. In time, these pieces will become the rocks and mountains of the future.

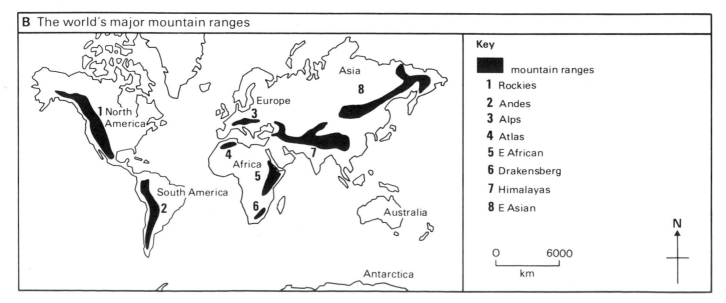

B The world's major mountain ranges

Key

■ mountain ranges
1 Rockies
2 Andes
3 Alps
4 Atlas
5 E African
6 Drakensberg
7 Himalayas
8 E Asian

0   6000
km

N

1 Make a copy of map **B** to show the world's major mountain ranges. Remember to include these things:
   – a frame
   – a title
   – a key
   – a scale
   – directions

2 Activities in the following list are sometimes difficult in mountain areas. For each activity, explain what the problems are. The letters refer to clues on sketch **A**.
   a Farming in a mountain valley (K,L,M,O)
   b Driving a car (J,H,K)
   c Farming on mountain slopes (I, G, K,H)
   d Living in a mountain village (N, K, M)
   e Climbing mountain peaks (A, E, B, C, F)

3 Contour lines are lines drawn on a map to show how high the land is. Map **D** shows what a mountain area looks like when contour lines are drawn. Chart **E** (right) shows some common contour shapes and what each would look like as a landform.
   a Copy chart **E**. Draw the contour shapes and sketches and write out the information.

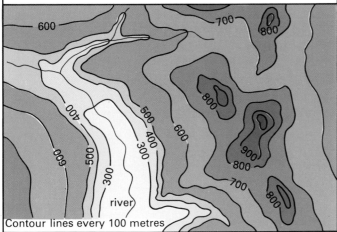

D Contour map of a mountain area

600 700 800 400 500 400 300 600 800 900 800 700 800 600 500 300

river

Contour lines every 100 metres

b Lay a sheet of tracing paper over map **D**. Print in the following letters where you think they should be:

Letter	Description
M	mountain top (five times)
V	valley (two times)
S	steep slope (two times)
G	gentle slope (four times)

E

Sketch of feature	Shape of contour	Definition	Description
		hill or mountain top	contours in circles
	river	valley	fingers pointing upstream
		steep slope	contours close together
		gentle slope	contours far apart

4 Write a paragraph to explain why each of the following happens in mountain areas. Refer to diagrams **A** and **C** in your answer.
   a landslides
   b spring floods
   c snow avalanches

Advanced work
5 In sketch **F** (below) several good uses have been found for a mountain area.
   Use the following headings to say how a mountain area can be used nowadays: water supply, holidays, electricity, stone for building, growing trees for timber.

Summary
Mountain areas are very difficult places to live in. Nature's work of breaking down mountains can make life dangerous for man. Today, mountain areas can be useful in some special ways.

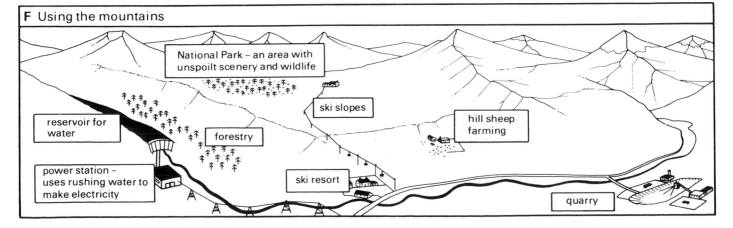

F Using the mountains

National Park – an area with unspoilt scenery and wildlife

ski slopes

reservoir for water

forestry

hill sheep farming

power station – uses rushing water to make electricity

ski resort

quarry

Living in the desert

Deserts are places where there is not enough water for people, animals, or crops. Deserts have always been places to avoid. An Atlas map of where people live shows that very few live in desert areas.

A World deserts

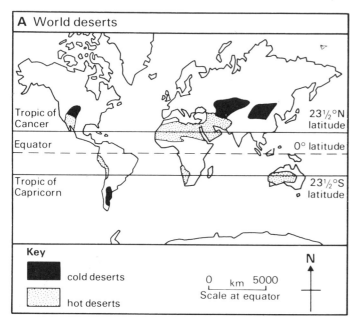

Key

■ cold deserts

▒ hot deserts

0   km   5000
Scale at equator

N

Tropic of Cancer .... 23½°N latitude

Equator .... 0° latitude

Tropic of Capricorn .... 23½°S latitude

Map **A** shows that there are different kinds of desert. Some are the hottest places on earth, but some are quite cold. Hot deserts are found near the tropics. Cold deserts are found in the centre of some large continents.

Deserts do have a little rain, but some years are completely dry. Rainwater either sinks into the ground or quickly dries up under the hot sun.

Deserts are not completely deserted. Some people live in oases where there are **springs**. A spring is a waterhole where water from underground rocks comes to the surface. Some groups of people wander from oasis to oasis with camels and sheep. They live off what their animals provide and by trading.

Photograph **B** is part of a typical oasis in the Sahara desert. Almost anything can be grown as long as there is water. Tall palm trees shade the plants which do not like too much heat. Life in the oasis depends on the water. If this dries up, the people have to move away.

Some deserts are better known for oil than for date palms and camels. There is oil in rocks deep beneath some of the countries of the Middle East. Living and working in the desert is difficult, but oil is so valuable that these problems must be overcome (photograph **D**).

Camps to house the workers have been built. The houses must have air-conditioning units to keep them cool inside. Water comes from bore holes sunk into the rocks. When drilling has finished at one place, everyone moves on to a new site somewhere else.

Some countries dream of turning their desert lands into good farmland. They sell oil to other countries, then spend the money on drilling for underground water. With the water, they can **irrigate** (water) the land and grow more food. This makes the future more secure for everyone.

C Rocks beneath the desert

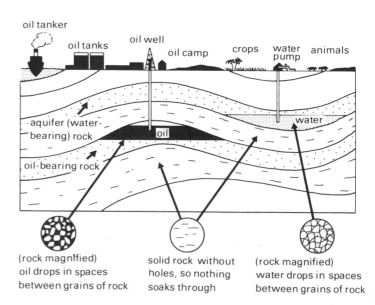

oil tanker

oil tanks   oil well   oil camp   crops   water pump   animals

aquifer (water-bearing) rock

oil

water

oil-bearing rock

(rock magnified) oil drops in spaces between grains of rock

solid rock without holes, so nothing soaks through

(rock magnified) water drops in spaces between grains of rock

1 Write down the sentence from the opposite page which says what a desert is.

2 Draw a map of the world to show the hot desert areas. Label in these desert names in the right place:

Name	Location
Sahara	North Africa
Namib	Southern Africa
Arizona ⎤	North America
Mohave ⎦	
Atacama	South America
Great Australian	Australia
Arabian	South West Asia

3 Make a sketch of photograph **B**. Label these typical oasis features on the sketch: date palms, crops, buildings, sand dunes.

4 a The Middle East countries listed below have been jumbled up by putting the wrong endings on each. Use an Atlas to sort out the correct names for these countries:

Middle East jumble

Saudi Ar<u>aq</u>	Jor<u>abia</u>
Ir<u>wait</u>	Is<u>ypt</u>
Ku<u>dan</u>	Eg<u>rael</u>

b Imagine you are working at an oil rig in a Middle East desert. Write a letter home to describe the area, and say what kind of problems you face every day.

D

E Rocks and rock terms

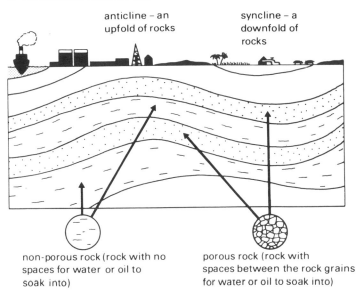

anticline – an upfold of rocks

syncline – a downfold of rocks

non-porous rock (rock with no spaces for water or oil to soak into)

porous rock (rock with spaces between the rock grains for water or oil to soak into)

Advanced work

5 Diagram **E** show the proper terms used to describe rocks and rock layers. Compare this diagram with diagram **C** on the opposite page, then complete the following sentences, choosing one of each pair of words:

a Oil is found in $\dfrac{\text{porous}}{\text{non-porous}}$ rocks.

b Oil becomes trapped in $\dfrac{\text{a syncline.}}{\text{an anticline}}$

c Water is found in $\dfrac{\text{porous}}{\text{non-porous}}$ rocks.

d Water underground collects in $\dfrac{\text{anticlines}}{\text{synclines}}$.

> **Interesting fact:**
> The highest recorded temperature in the shade is 57.7°C at Al Aziziyan in Libya on 13 September 1922.

Summary
Deserts are places where there is not much water. People can live by farming only at oases or where water is piped in. The search for oil also brings people to live in the desert, even if only for a short time.

Only one-fifth of the surface of the Earth is used by people. The rest is too dry, too cold or too steep. Nearly all of this useful one-fifth is farmland, with its people living on farms or in villages. Two out of every three people on earth live in farming families. The other third live in towns and cities and earn a living from making things, or doing things for other people.

As the cartoon **A** shows, old-fashioned, **traditional** farming needs a lot of hard work by a lot of people. Most farming is still done like this. In rich countries machines have replaced people and fewer farm workers are needed.

Map **B** (below) shows where some of the main types of farming are found.

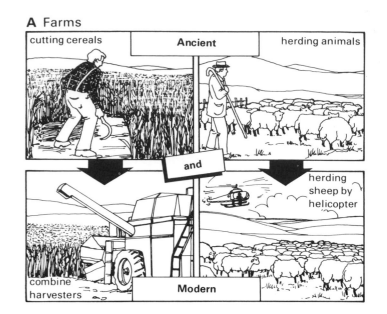

A Farms

cutting cereals | **Ancient** | herding animals

and

combine harvesters | **Modern** | herding sheep by helicopter

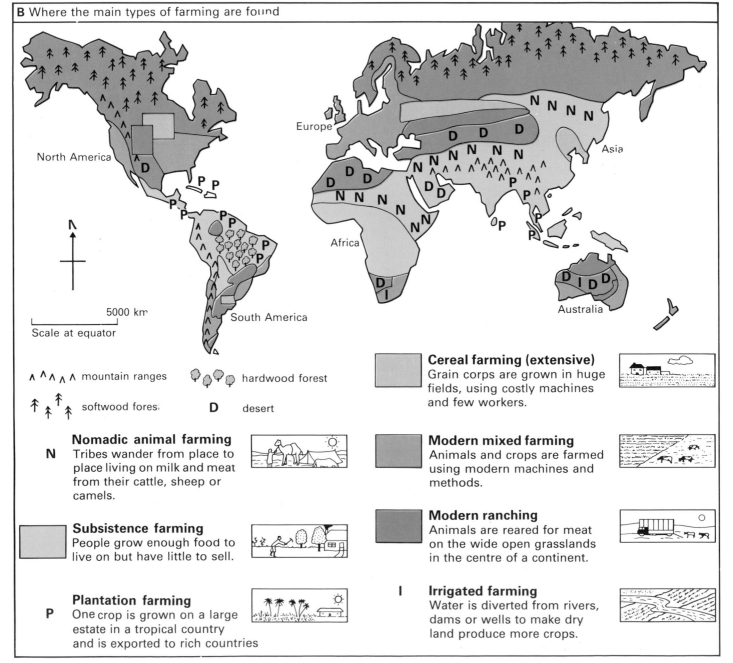

B Where the main types of farming are found

North America
Europe
Asia
Africa
South America
Australia

5000 km
Scale at equator

∧ ∧ ∧ ∧ ∧ mountain ranges

🌳🌳🌳 hardwood forest

🌲🌲🌲 softwood forest

D desert

N **Nomadic animal farming**
Tribes wander from place to place living on milk and meat from their cattle, sheep or camels.

Subsistence farming
People grow enough food to live on but have little to sell.

P **Plantation farming**
One crop is grown on a large estate in a tropical country and is exported to rich countries

Cereal farming (extensive)
Grain corps are grown in huge fields, using costly machines and few workers.

Modern mixed farming
Animals and crops are farmed using modern machines and methods.

Modern ranching
Animals are reared for meat on the wide open grasslands in the centre of a continent.

I **Irrigated farming**
Water is diverted from rivers, dams or wells to make dry land produce more crops.

1 Trace the map opposite and copy the definitions as part of the key.

2 Write one sentence to name the main type of farming found in each of these places.
   a Western Europe
   b Northern Australia
   c The USA/Canadian border in middle of North America
   d The 'Horn' of Africa (North East Africa)
   e The island of Sri Lanka, south of India
   The list below may help you.

Types of farming
Here they are – but not in this order:
Nomadic animal farming
Extensive cereal farming
Plantation farming
Modern ranching
Mixed farming

3 Make a sketch of photograph **C** (below), and use the clues to write a paragraph of description of this way of life. Use the part-sketch **D** to help you.
   Clues
   Afghanistan, Bactrian camel, tent, milk, firewood, travel, grass for feeding, plains.

C

D Part-sketch of photograph C

4 Draw a cartoon to show an example of how old-fashioned farming has changed to modern. Make it simple but neat. As an example, cartoon **E** shows horses being replaced by tractors.

E Old and new farming methods

5 Choose any one continent. Imagine you have travelled across it, working for a short time on a number of different farms. Write an account of your travels, saying what you thought of the different ways of life. Use the map opposite and your Atlas to mention real places which you might have passed through. The verse below, written by the famous American song-writer, Woody Guthrie, may help you.

'California, Arizona, I've gathered your crops,
Then on up to Oregon to help with the hops,
Beet from the ground and grapes from the vine,
To put on your table that white sparkling wine.'
(From 'Pastures of Plenty' by Woody Guthrie)

Summary
Most of the world's people live on the land. Traditional ways of farming need a lot of people. In rich countries few people work on the land. Machines have replaced them.

Unit 2.2
Nomads

Nomads are people who roam from place to place, usually with herds of animals. The Boran tribe of East Africa are a nomadic people. They move about in large family groups. Each family may have up to one hundred men, women and children.

Most Boran people still live in the traditional way. They live in a **manyatta**, which is a village of huts made of branches and mud. Every few years the family leader decides to move the manyatta to a new place. Because they live in an almost treeless desert, the people take their hut branches with them on their camels and donkeys. Map **A** (below) shows the different places where one group has lived since 1950. The list beside the map shows the ages of the people in the group, and how many animals they have.

All the family's wealth is in their animals. Boran people live mainly on milk and some meat. Their

clothes are made from animal skin, or from cotton bought from animal sales. One sheep can be exchanged for 10 m of cotton cloth. In 1980 other exchange rates were:
> 1 camel = 2 cows = 30 goats or sheep = 2 donkeys
> 2 small elephant tusks = 1 camel = £100

The animals are tended by young men and children. The herds of sheep, goats and cattle are taken away from the manyatta to eat grass in parts of the desert where rain showers have caused grass to grow. A special look-out is kept for rain clouds. The herders walk the animals towards rain clouds in the hope of finding fresh grass.

The young men often camp with young men from other Boran families and set up a camp called a **boma** in the desert. They must return to their family manyatta in June when there is a family prayer ceremony to celebrate the passing of the year.

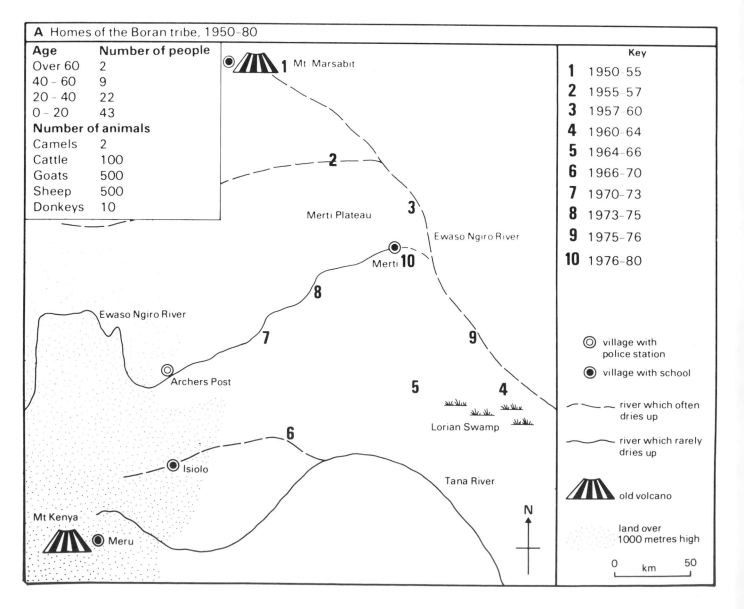

A Homes of the Boran tribe, 1950–80

Age	Number of people
Over 60	2
40 – 60	9
20 – 40	22
0 – 20	43
Number of animals	
Camels	2
Cattle	100
Goats	500
Sheep	500
Donkeys	10

Key
1	1950-55
2	1955-57
3	1957-60
4	1960-64
5	1964-66
6	1966-70
7	1970-73
8	1973-75
9	1975-76
10	1976-80

Mt Marsabit, Merti Plateau, Ewaso Ngiro River, Merti, Ewaso Ngiro River, Archers Post, Lorian Swamp, Isiolo, Tana River, Mt Kenya, Meru

◎ village with police station
◉ village with school
river which often dries up
river which rarely dries up
old volcano
land over 1000 metres high

N

0 km 50

1 Study the page opposite.

2 Answer these questions with a sentence for each answer.

  a What is the name of the **volcano** in the south-west corner of the map?

  b Which volcano did the family live near in 1952?

  c Which village did they live near in 1980?

  d Name the nearest village with a police station.

  e Which river flows (sometimes) past the family manyatta?

3 Look at photo **B** and copy sketch **C**.

B A Boran manyatta

C Sketch of photograph **B**

4 The Merti region receives less than 250 mm of rain a year, as shown in the graph below. It is a desert region. Copy the graph and write fifty words altogether to answer these questions:

  i Which months are rainy and which are not?

  ii How is their climate different from ours?

  iii When do you think the rivers flow?

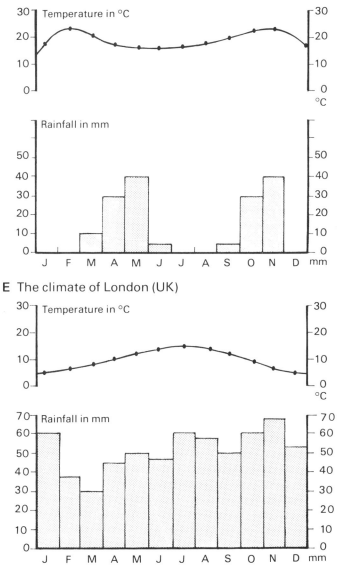

D The climate of Merti (Kenya)

E The climate of London (UK)

5 Write an account of this Boran clan using the information in the lists, maps and photographs on these two pages. Here are some words which you might use: work, ages, wealth in pounds, school, walking, British gypsies.

Summary
Nomads are wanderers. The Boran tribe move their animals from pasture to pasture in the desert of Northern Kenya.

Unit 2.3
Subsistence farming

Subsistence farming is farming which only just feeds the farmer's family, with little or no food left over to sell at the local market. Subsistence farmers have little money to buy things. Their whole family life depends on the crops in their fields. A long dry spell or a flood may turn a poor family into a starving one, as we sometimes hear on the News. Table **A** on the right shows the types of food grown in two areas.

Some subsistence farmers wear out the soil in one place, then move on to another. They burn the forest, plant their crops and build new villages. When their crops begin to get poorer, the farmers look for somewhere else. This farming is called **shifting cultivation**, and needs a lot of land to feed a few people. In parts of South-east Asia families have farmed the same patch of land for generations. They keep the soil in good condition by working hard to mix animal manure into it.

Most subsistence farmers have to face the fact that, with improved medicine, there are more mouths to feed from the same land. There is no easy way to do this, as any improvements need new skills and equipment and both of these cost money!

Photograph **B** shows a forest in Nigeria, a country in West Africa. Farmers clear this forest by fire each year in the dry season from

A Types of food grown

	West Africa	Indonesia
Grain	millet and maize	rice, tapioca
Root crops	yams and ground-nuts.	ground nuts
Vegetables	cucumber, red beans, red peppers	soya beans, peppers
Fruit	many tropical fruits	many tropical fruits
Meat	goats, chickens	chickens

December to March. The farmer's main crop is yams, a root crop. The busiest times on his new land are in April at the start of the rains when he plants his yams, and October at the end of the rains when he harvests them. A yam is like a potato but much bigger. Though he farms new land each year he keeps just one house, in his home village.

Photograph **C** shows an Indonesian farmer tending his rice crop. Rice is the main crop of South-east Asia. Here in Java near the volcanic mountains the soil is thick. Rainfall is so plentiful (150 cm in the wettest season from October to April) and the weather is so warm that the farmer is able to grow three separate crops of rice each year. His family has lived in the same village and has worked the same piece of land for as long as anyone can remember. Hard work is needed all through the year to keep the farm in good condition. This type of farming, where such care is taken of the land, is called **intensive cultivation**.

Rice needs water and sunshine. It is said that 'rice grows with its feet in water and its head in the sun'.

B

C

16

Things to do

1 Write an account of what subsistence farming is. Use the opposite page and diagram **D** to help you.

D Subsistence farming life

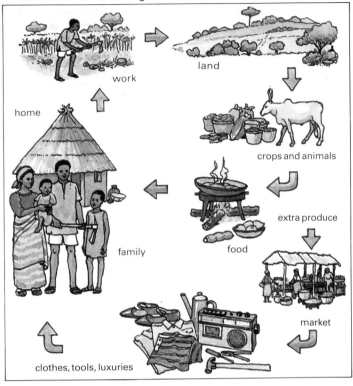

work
land
home
crops and animals
family
food
extra produce
market
clothes, tools, luxuries

2 Write a sentence to explain or describe each of these:
a shifting cultivation
b yams
c intensive cultivation
d three crops a year
e subsistence farming

3 Draw a sketch of the photograph of an Indonesian farm (photo **C**). Use the part-sketch **E** to help you. Add on five labels.

E Part-sketch of photograph **C**

4 Copy graph **F**. Use it to plot a scatter graph to show this information:

Place	Rainfall (mm per year)	Rice harvest (tonnes per hectare)
1	700	1.4
2	2000	2.0
3	500	1.0
4	800	1.2
5	1500	1.6
6	600	0.8
7	1750	1.8
8	1250	1.2
9	1000	1.4
10	400	0.4

Look at your finished graph. Copy out this sentence and choose the right words from the brackets:
The (more/less) the rainfall the (bigger/smaller) the rice harvest.

F Scatter graph plotting rainfall against rice harvest

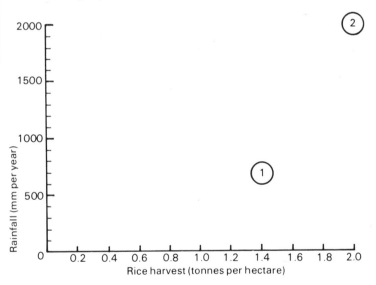

5 Imagine you are one of the farmers on the previous page. Write a letter to the other one explaining your year's work and any interesting things that happened.

Summary
Subsistence farmers and their families live from the food they produce. They have little left over to buy anything.

17

Unit 2.4
Plantations and irrigation

Two interesting ways of farming in hot countries are described on this page. They are plantation farming and irrigation farming. By these two methods farmers are able to produce more food than by subsistence farming.

Plantations are farm estates where the 'planter' grows one crop for cash, rather than for food. Plantations are usually found in tropical countries and their produce is exported to the rich northern countries. Map **A** shows where most plantations are found. Many were started between 1700 and 1900 by European people in their **tropical colonies**, especially in South-east Asia. This was because this region had a lot of people who were prepared to work hard for low pay. In the West Indies and the Americas the workers were often slaves. Wages are still low on many plantations, but other jobs are scarce.

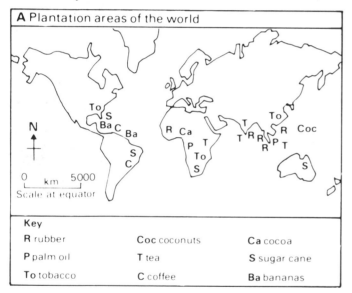

A Plantation areas of the world

Key

R rubber	**Coc** coconuts	**Ca** cocoa
P palm oil	**T** tea	**S** sugar cane
To tobacco	**C** coffee	**Ba** bananas

The layout of a plantation is usually neat. It has a central collecting place where workers live, and roads to all parts of the estate. Some plantation managers were said to 'rob' the soil. They got as much as they could from the land and left it spoiled and useless. Most plantations, however, are run well and use expert methods. A big problem is the world price for a crop. If it suddenly goes down, the plantation owner may be ruined.

Irrigation means supplying water to dry land. It allows plants to thrive in places which would be unproductive if left to nature. The water may be obtained from springs or wells, or rivers. Sketch **B** shows some of the ways in which water is taken to the dry fields.

Irrigation is a great invention. It helped to build up many other sciences in Ancient Egypt where it was first practised. For example:
Astronomy – to predict the dates of the River Nile floods.
Geometry – to mark out the fields after flooding.
Engineering – to make water-moving machines.

In ancient times the irrigated areas of Egypt near the River Nile, and India and Pakistan near the River Indus, were the richest in the world. Another famous area was Mesopotamia – between the rivers Tigris and Euphrates in Iraq and Syria.

Nowadays great dams are built to hold back flood water and to channel it onto fields. Such schemes cost millions of pounds. Common crops grown under irrigation are rice, dates, sugar cane, cotton, citrus fruits such as oranges, lemons, grapefruit, and vegetables such as tomatoes, onions and lettuce.

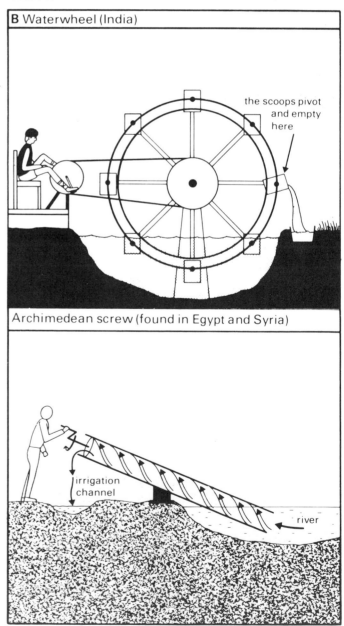

B Waterwheel (India)

the scoops pivot and empty here

Archimedean screw (found in Egypt and Syria)

irrigation channel

river

1 a Draw map **A** showing the main plantation areas.
 b Draw the articles in figure **C** and label each one to say which plantation crop it came from.

C

2 Write one sentence to answer each of these questions:
 a What is plantation farming?
 b What is irrigation farming?
 c How does the Archimedean screw work?
 d Which plantation crop is grown in Australia?
 e Which plantation crop is grown on islands in the Pacific Ocean?

3 Draw a sketch of photograph **D** (below) of a scene in Syria. Copy the part-sketch **E** and add your own labels.
 Write a paragraph to explain what is happening in the photograph.

D

E Part-sketch of photograph D

4 Draw a diagram to show what you think is happening in photograph **F**. It shows a scene in India.

5 Invent your own plantation. Draw a plan of it, using diagram **G**. Label in storage barns, houses, fields, roads and trucks. Say where in the world it is, and what it produces.

G Simple plan of a plantation

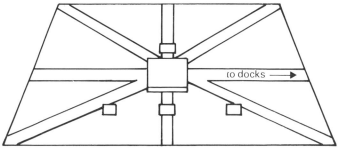

to docks →

Summary
Plantations are farm estates where one cash crop is grown, usually in tropical countries. Irrigation farming adds water to the land. Both types of farming get more from the land than subsistence farming.

Modern farming/The plains of grain

Photograph **A** shows a grain storage **elevator** at Avonmouth, UK. Grain is the seed from special types of grass (see sketch **B**) and it is most people's main food. It is moved by train and ship from countries which produce more than local people can eat (grain selling areas) to those which want it, can pay for it, but cannot grow enough of it (grain buying areas). Grain storage elevators like these at ports are a reminder that grain is second only to crude oil as a **commodity**. A commodity is anything which is traded in its raw state.

All grains need rain and sun, but some need more than others. This table shows the five main types of grain, the conditions they need and the main areas where they grow:

Type	Growing conditions	Main growing areas
Wheat	Warm, dryish	USA, N China, India, USSR
Rice	Hot, wet	S E Asia
Maize (corn)	Warm, wet	USA, USSR, India
Barley	Cool, dryish	USA, Canada, W Europe
Sorghum	Hot, dry	Africa, S Asia

Map **B** shows that the world's main grain selling areas are the vast rolling lowlands of North America and Australia, and the flat 'pampas' of Argentina. In these areas, grain is grown in empty prairies, where huge fields of wheat and barley stretch as far as you can see. There are few, if any, buildings to be seen. The few workers on each

A

grain elevator

farm are usually highly skilled as farmers and also as mechanics, so they can drive and service the tractors and harvesters which are used. This sort of cereal farming is called **extensive** farming.

Strangely, less wheat is grown in Canada than in India. However, Canada has only 20 million people to India's 500 million. This means that Canada has too much grain (a **surplus**) and India has too little grain (a **deficit**), so Canada sells grain to India.

The main buyers of grain are the densely populated countries of North-western Europe. Other countries like India and Bangladesh want the grain but cannot afford to buy very much. This sometimes leads to the peculiar state of affairs in which surplus grain is destroyed in North America while people go hungry in Africa and Asia. The costs of transport would be too much even if the grain were given away, so it is burned.

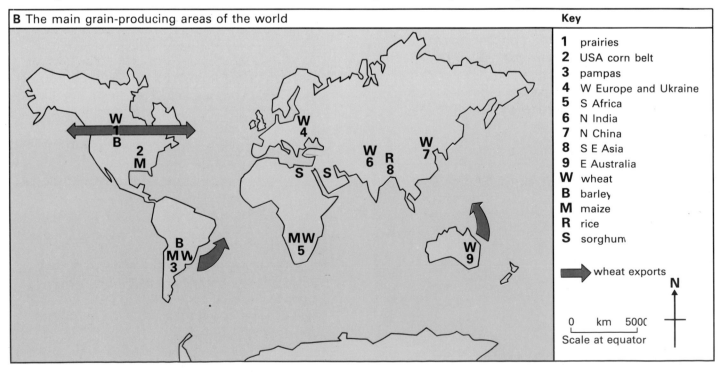

B The main grain-producing areas of the world

Key

1 prairies
2 USA corn belt
3 pampas
4 W Europe and Ukraine
5 S Africa
6 N India
7 N China
8 S E Asia
9 E Australia
W wheat
B barley
M maize
R rice
S sorghum

→ wheat exports

0 km 5000
Scale at equator

N

C What different types of grain look like

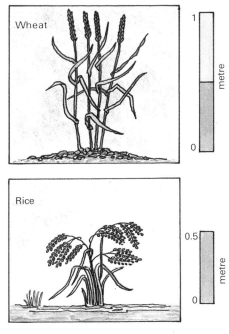

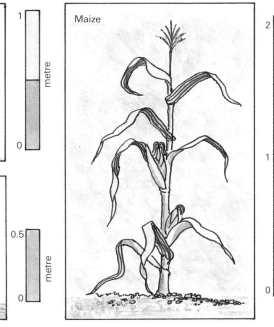

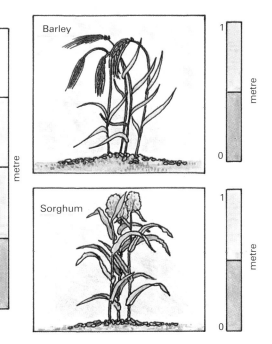

1 Study diagram **C**. Draw each of the five types of grain, putting in the 1 metre scale and the figure of a person on each drawing.

2 Write a sentence to answer each of these questions:
a Which grain crop can grow in hot dry conditions?
b Which is the main grain crop in South-east Asia?
c Which is the main grain crop in Northern China?
d Which grain plant is tallest?
e Which commodity is traded in greater quantities than grain?

3 Table **D** shows the total amount of different cereals grown annually. Put the list into order: highest output in first place, lowest output in eighth place. Write a paragraph to say which grains are most important.

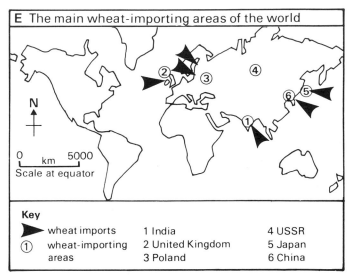

E The main wheat-importing areas of the world

Key
➤ wheat imports
① wheat-importing areas

1 India	4 USSR
2 United Kingdom	5 Japan
3 Poland	6 China

4 Draw **one** world map to show the main wheat-producing areas, the main wheat-importing areas, and the main directions in which wheat is traded. Use maps **B** and **E** to help you.

5 Write a short paragraph to explain each of these terms, and give an example:
a grain surplus area   **d** extensive farming
b grain deficit area   **e** burning of grain
c commodity

D   World cereal output (1980)

Cereal type	Main producer	World output in millions of tonnes
Barley	USSR	180
Maize	USA	350
Millet	India	35
Sorghum	USA	60
Oats	USSR	50
Rice	China	425
Rye	USSR	30
Wheat	USSR	490

Summary
Cereals are grain from special types of grass such as wheat and rice. They are very important foods. They are traded round the world as commodities.

Unit 3.2
Animal farming

Many of the everyday things around us come from farm animals. We eat their meat, drink their milk and wear their wool and skins. In poor countries families keep their own animals. In rich countries we leave this job to skilled farmers who may use **scientific methods**.

In rich countries animals are reared as a way of making money. This is not the view of many people in Africa and Asia. In India cows are sacred and no one is supposed to harm them. In many parts of Africa a man shows how rich he is by how many cattle he owns. His cattle are a status symbol like a rich man's car in Europe or the USA. (See cartoon **A**.)

What you eat is often a guide to how wealthy you are. People in rich countries eat a lot of meat. People in poor countries eat little meat. Some animals are kept for other products. For example, most sheep in Australia are reared for their wool.

In most parts of the world animals are allowed to range over the grazing land with little attention. In rich countries in North America and Europe farmers make certain the grass is as rich as it can be. They grow **fodder** crops like hay and turnips to feed their animals in the winter. Dairy cattle are milked by machine. Some animals are reared entirely indoors, in pens. This is called **factory farming**. Some of the main animal-rearing areas are shown on map **B**.

Ten times as much land is needed to produce a tonne of beef than is needed to produce a tonne of grain. Some people say that if we ate more grain and less meat there would be plenty of food for everyone on Earth.

A Status

C Cows or ploughs?

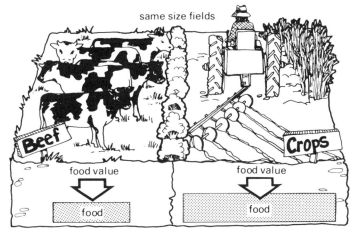

B The main animal-rearing areas of the world

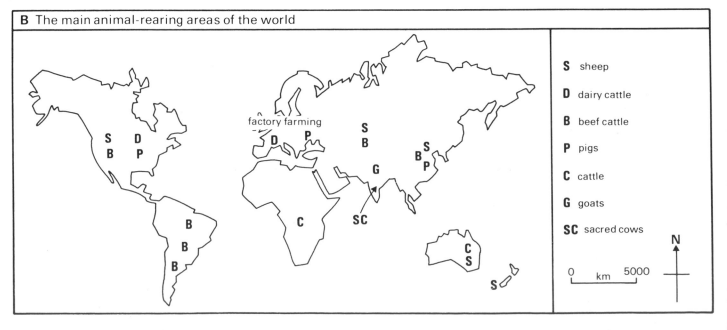

S   sheep

D   dairy cattle

B   beef cattle

P   pigs

C   cattle

G   goats

SC   sacred cows

1 To each of these 'heads' put the correct 'tail'.

a Wealthy people often   are status symbols.

b Many African cattle   are fed to animals in winter.

c Indian cattle   live mostly indoors.

d Fodder crops   eat a lot of meat.

e Factory farm animals   are sacred.

2 Draw the outline map **B** showing where the main areas of animal rearing are.

3 Think of your father or mother dressed up to go out somewhere. (Surely they go out sometimes!) Draw a simple sketch of them with labels showing what different parts of their clothing might have originally been and which countries they may have come from.

4 **a** Put the list (**D**) of the top five animals and animal foods into order, and write it down in your books.

D The top five farm animals and animal foods

Animal food products (millions of tonnes)		Animals (millions)	
Beef	45	Goats	480
Hen eggs	25	Sheep	1200
Pork	40	Cattle	1200
Cow's milk	440	Chickens	1300
Chicken meat	10	Pigs	750

b How could keeping computer records help a farmer? Cartoon **E** shows what they might be like!

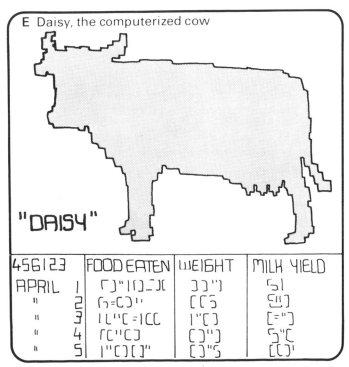

E Daisy, the computerized cow

"DAISY"

456123	FOOD EATEN	WEIGHT	MILK YIELD
APRIL 1			
" 2			
" 3			
" 4			
" 5			

5 Write an imaginative essay about how different people feel about animals and the places where they live. Use the opposite page and your own knowledge of people and places to get ideas. Cartoon **F** (below) may help you.

F

Summary
We get many everyday things from animals. In rich countries, animals are reared to make money. In some poor countries they are status symbols.

Modern mixed farming

The two farming areas on list **A** are two of the best farming areas on earth. The **Corn Belt** produces a quarter of the world's maize and a fifth of the world's beef and pork. Western Europe has great surpluses of dairy products and of some vegetables, though the high standards of living make imports of beef, wheat and tropical crops necessary.

A

Corn Belt States in the USA	Chief farming countries of Western Europe
Wisconsin (South)	France
Michigan (South)	West Germany
Illinois	Southern England
Indiana	The Netherlands
Ohio	Northern Italy
Missouri	Belgium
Kentucky	Denmark

Both areas have mainly **mixed farming**. This means that livestock and crops are found on most farms. The idea is that some of the farm crops are grown to feed the livestock. The livestock droppings make the soil richer, ready for the next year's crop. In other words, one helps the other. Diagram **B** shows how this works on a typical farm in the Corn Belt. The Corn Belt gets its name because so much corn (maize) is grown there. Most of the farms have pigs, cattle and maize.

Farms in Western Europe are all different from each other. Each farm has its own special conditions of weather, slopes and soil, and each farmer has his own ways of doing things. Most farms, however, have something in common. A typical mixed farm in Europe will grow wheat or barley, with potatoes and fodder crops such as kale or turnips. It will have a small herd of dairy cattle for milk or bullocks for slaughter. Most farmers keep several pigs to dispose of waste food and to provide pork for the family table. A typical farm is shown in sketch **C**.

Even though Western Europe and the Corn Belt are far apart, they use similar methods. Here are some of them:
1 **Use of all products.** Little is wasted. For example, surplus apples are used to feed pigs.
2 **Crop rotation.** Different crops use different minerals in the soil. If crops are 'rotated' the soil does not become exhausted. See diagram **D**.
3 **Mechanical aids.** Few people work on the land in these two areas. Wherever possible machines are used to replace unskilled workers.
4 **Heavy use of fertilizers.** To keep the soil producing at peak output, manure and chemicals such as ammonium nitrate and calcium phosphate are mixed into it. Huge factories supply millions of tons of such chemicals each year.

B How maize and meat help each other

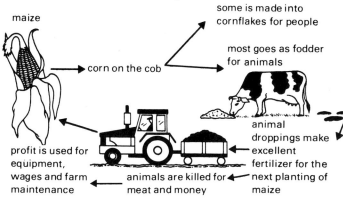

maize

some is made into cornflakes for people

corn on the cob

most goes as fodder for animals

animal droppings make excellent fertilizer for the next planting of maize

profit is used for equipment, wages and farm maintenance

animals are killed for meat and money

D Crop rotation

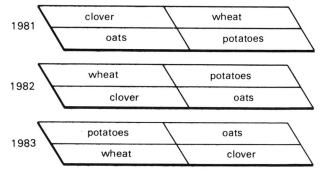

1981	clover	wheat
	oats	potatoes

1982	wheat	potatoes
	clover	oats

1983	potatoes	oats
	wheat	clover

C Typical mixed farm in Western Europe

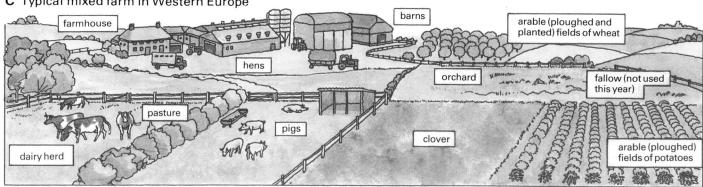

farmhouse

barns

arable (ploughed and planted) fields of wheat

hens

orchard

fallow (not used this year)

pasture

pigs

clover

dairy herd

arable (ploughed) fields of potatoes

1 Copy diagram **B**.

2 Copy out and fill in crossword **E**.

E Crossword

Clues
a <u>A</u>m<u>aiz</u>ing breakfast cereal.
b Animal droppings are one kind of _____.
c Crops going round in circles?
d State in the South-west of the Corn Belt.
e A soft life for one crop. It is in _____.
The correct answers are in this list: fertilizer, rotation, cornflakes, clover, Missouri.

3 Make a list of the scientific methods which make mixed farming such a success in Western Europe and the Corn Belt.

4 Study farm plan **F**. This farm grows potatoes, maize, grass and wheat, using crop rotation. Make three copies of plan **F** in your book. Mark on them how you would plant the crops in 1982, 1983 and 1984. Make sure no crop is in the same field two years running.

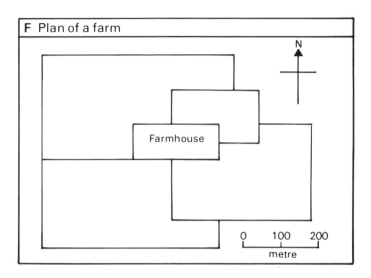

F Plan of a farm

Farmhouse

0   100   200
metre

5 Copy the simple map of the world (**G**).
   a Use your Atlas and list **A** of states and countries to shade the area of the Corn Belt and Western Europe on this map.
   b Draw a simple map of the USA to show the Corn Belt. Which Corn Belt city lies on the southern end of Lake Michigan?

Summary
The Corn Belt and Western Europe are rich farming areas. Their farms have both livestock and crops. This is mixed farming.

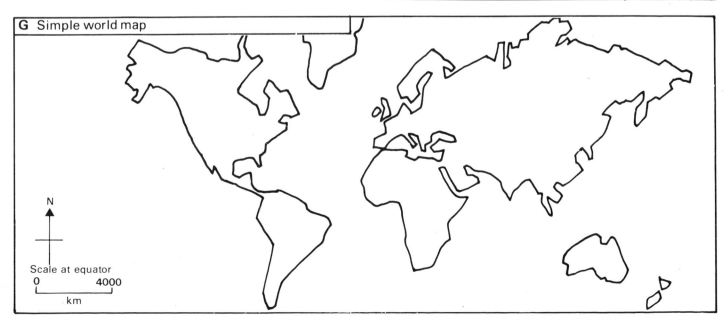

G Simple world map

N

Scale at equator
0   4000
km

25

Unit 3.4
Ideas for the future

The number of people in the world is likely to double in the next fifty years. We will need new farming methods to feed this population of 6 000 000 000 people. Some farmers are already using new methods to increase production. Here are some of them:

1 **Drip irrigation** (see sketch **A**). Plastic pipes are laid on fields. The pipes have holes every few centimetres. Crop seeds are planted near the holes. Water (from wells or streams) is allowed to flow through the pipes and to drip through the holes to water the crop. Very little of the water is wasted, so this is a cheap way of irrigating dry farmland. Even salty water works well like this. This method is already found in the USA, Australia, Israel, South Africa and Mexico. It is used to grow vegetables like lettuce and cucumber.

2 **Farming under polythene** (see sketch **B**). When plants are covered with thin polythene sheets they keep warm and moist, and grow faster. This is now done in Britain, Europe, Japan and the USA.

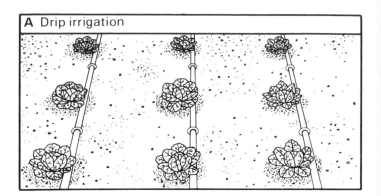

A Drip irrigation

B Polythene sheets covering crops

3 **Superseeds** (see cartoon **C**). By crossing different breeds of plants, new strains of wheat, rice and maize have been made. They grow faster and fight off disease better than old strains. In South-east Asia they have already increased rice harvests where they have been used, though care has to be taken not to wear out the soil. A lot of fertilizer has to be used to keep the soil healthy.

4 **Fertilizer** (see diagram **D**). In India, animal dung is dried and burned for heat. A new method uses the dung to provide gas for heating and lighting houses (without smell!) and provides liquid manure for the fields. Dung from five animals provides heat, light and fertilizer for one family, and the method is cheap. Simple ideas like this are the best ones. They use common things but in better ways than before.

5 **Food storage** (see cartoon **E**). One tonne of food in every five tonnes grown in poor countries is ruined in storage. It gets damp and rots, or is eaten by mice. Simple food storage bins are now being made from local materials all over Africa to keep grain in good condition.

E

Fat rats = poor farmer

Food Storage

A dead loss for rats = rich farmer

The big problem with all these methods is persuading farmers to use them instead of the old methods they are used to. Old customs die hard.

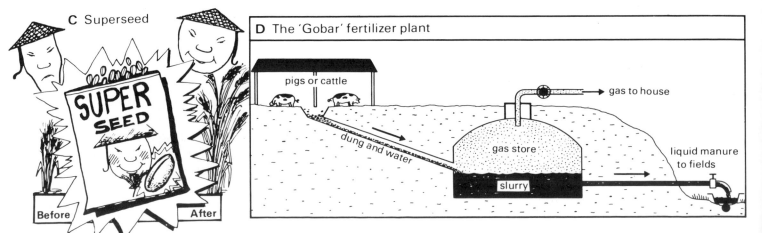

C Superseed

SUPER SEED

Before   After

D The 'Gobar' fertilizer plant

pigs or cattle

dung and water

gas to house

gas store

slurry

liquid manure to fields

1 Which of the new farming methods on the opposite page seems to you to be the most interesting or the most useful? Copy out the sketch or diagram of that method.

2 Write a sentence to say which modern method would include
a shovelling dung
b putting holes in pipes
c building huge baskets on legs
d breeding plants
e farming under the sheets

3 Copy the part-graphs **F** and finish them off from this information:

Year	Tonnes of wheat/hectare
1979	1.65
1978	1.60
1977	1.55
1976	1.50
1975	1.47
1974	1.45
1973	1.46
1972	1.47
1971	1.55
1970	1.53

Write a paragraph to try to explain what the graphs tell you.

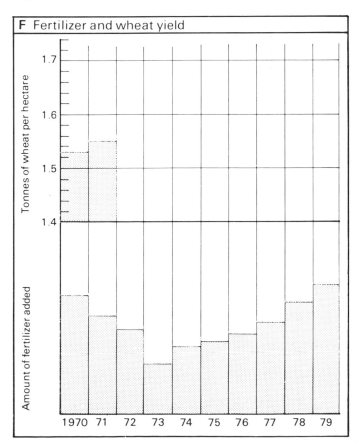

F Fertilizer and wheat yield

4 Irrigation is a problem in Bangladesh. Photograph **G** shows one solution. Draw a sketch of the photograph and write a paragraph to say what you think is happening.

5 Write a short conversation between an old farmer and his son, who is nagging his father to change his ways. The ideas in **H** may help you.

H

It's all too much trouble!

Your grandfather didn't have water pipes. Why should we?

I can't drive a tractor.

These seeds will make us enough profit to buy a tractor.

We need the animal dung for burning.

There's a mechanic's course in the village.

Summary
New ways of farming are being invented. They are needed to feed the growing number of people. Crop growing and crop storing can be improved.

Developing resources/What is a resource?

A **resource** is anything which people can use. It may be solid like iron ore, or liquid like crude oil, or a gas like natural gas from the bed of the North Sea. It may come from deep underground, like coal, or it may grow on the surface like pine wood. These are all raw materials. But resources may not even be materials. The skill of a **technician** or teacher is a resource! The more educated and skilful a nation's people, the greater its human resources are.

List **A** shows the top ten (by weight) resources produced in the world per year. Notice how the huge countries, the USSR, USA and China, are the biggest producers but not always the biggest exporters. In fact the USSR often has to import wheat and maize.

Resources are used to make things which people want, as diagram **B** shows.

The person who buys the thing is called a consumer. In this case the consumer buys paint to paint his house.

B Resource development

Some countries have few resources. South Africa is one of the fortunate regions with many. They are shown on map **C**. The only major mineral resource it does not have is oil.

South Africa has these advantages:
1 Vast mineral resources of coal and iron. These are essential to modern industry.
2 A huge gold output. Gold is used for currency (money) and jewellery. It gives South Africa the money to develop all her resources.
3 Skilled people. For 100 years South Africa has had a great number of people with industrial know-how.
4 A warm and moist climate (in the east), allowing farming.
5 An important position. Many ships sail into South African ports as they round the Cape of Good Hope.
6 Space. There is plenty of room for everyone. South Africa is a big country.

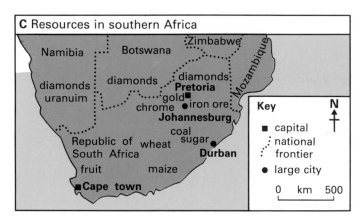

C Resources in southern Africa

If South Africa has a peaceful future it will become one of the richest countries on earth. It has a problem, though. White people are in power now. Sooner or later they will have to allow coloured and black people to share that power.

Peace is a resource.

A				
Resource	Annual amount produced (millions of tonnes)	Chief producer	Chief exporter	Uses
1 Coal and lignite	3500	USSR	USA	Fuel and chemicals
2 Crude oil	3000	USSR	Saudi Arabia	Fuel and chemicals
3 Iron ore	600	USSR	Sweden	Steel
4 Limestone	500	USSR	(Not exported)	Cement and chippings
5 Wheat	420	USSR	Canada	Food
6 Rice	400	China	Thailand	Food
7 Maize	350	USA	USA	Food and fodder
8 Potatoes	300	USSR	USSR	Food and chemicals
9 Barley	180	USSR	France	Brewing and food
10 Salt	150	USSR	USSR	Chemicals

1 Copy sketch **D** and the first paragraph on the opposite page.

D Some types of resource

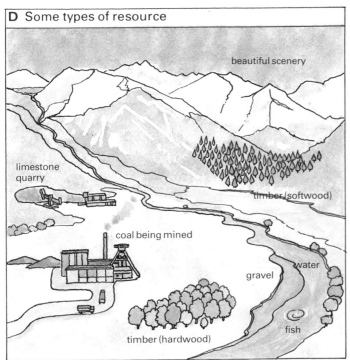

- beautiful scenery
- limestone quarry
- timber (softwood)
- coal being mined
- water
- gravel
- timber (hardwood)
- fish

2 Using only the words in diagram **E**, copy out and fill in these sentences.
  a A person who buys an item is a _____ .
  b Southern Africa's chief resource is _____ .
  c Something which is useful to people is a
  _____ .
  d The top producer of rice is _____ .
  e A very important human resource is _____ .

3 Copy the part-graph **F**. Using list **A**, fill in the resources in the correct order. Draw a bar to show the amount produced in 1980 and shade it to show the chief producer.

- gold
- skill
- consumer
- China
- resources

E

4 Use your Atlas to help you write a report on Southern Africa, using these headings:
  i   Countries
  ii Resources
  iii Main cities
  iv Location in the world (Draw a simple world map.)
  v Outlook for the future

5 How can 'peace' be a resource? Think about it and write fifty words or so to say what your ideas are. Then think about the following phrases. Are any of these things a resource?
  a spirit of teamwork
  a mild climate
  a position on a major trade route

Summary
A resource is anything which people can use. Resources have to be found and taken to factories. This is called resource development.

F Production of the world's top ten resources

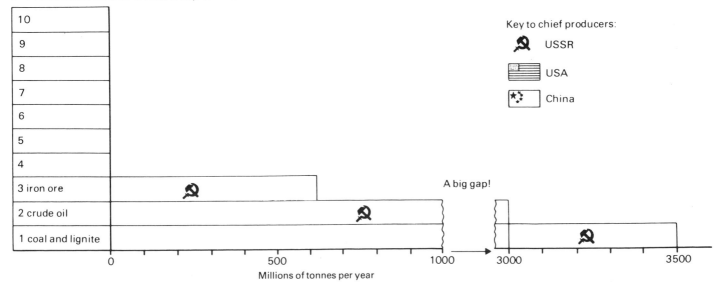

Key to chief producers:
- USSR
- USA
- China

10		
9		
8		
7		
6		
5		
4		
3 iron ore		
2 crude oil		
1 coal and lignite		

A big gap!

0   500   1000   3000   3500

Millions of tonnes per year

Developing resources in Mexico

A

New **oilfields** were discovered in Mexico during the 1970s. It was just like a pools win. Suddenly Mexico joined the first division of countries which have oil. Selling the oil all over the world is going to earn a lot of money for the Mexican people.

Map **B** shows where the main oilfields in Mexico are. Sometimes natural gas comes with the oil.

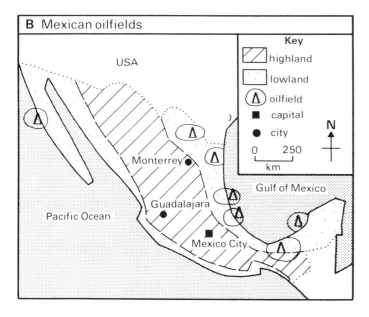

B Mexican oilfields

Key
- highland
- lowland
- oilfield
- capital
- city

USA

Monterrey

Guadalajara

Mexico City

Pacific Ocean

Gulf of Mexico

N

0   250
km

Most of the Mexican people are much poorer than people in the UK. Large areas of the country are mountainous and difficult to farm. The flat coastal areas are unhealthy because of jungle and swamps.

The capital, Mexico City, is getting bigger day by day. Poor people come to the city from country areas looking for work. There are no houses for them to live in, and not much work for them to do. Three-quarters of the city's 12 million people live in very bad conditions. About 2 million people live in shacks without a piped water supply. Three out of every ten are out of work. On top of all that, there will be twice as many people in Mexico in twenty years' time as there are now. Money from the sale of oil is going to be very useful.

The oil itself can be used to start up new industries which will mean more jobs. Fertilizers made from oil can help Mexican farmers grow more food. Plastics, textiles, and chemicals all come from oil. Crude oil is first **refined** in an **oil refinery**. This sorts out all the different chemicals which are in crude oil. Industries which use these chemicals are sometimes called **petrochemical industries**. The left-hand boxes in diagram **D** show some of the products of petrochemical industries.

The new factories can be run on power made by burning oil and gas.

Rich countries with money to invest are very interested in Mexico. They want to help develop Mexico's resources so that they can have a share in the money earned by the oil and the new industries.

Note: There is a problem with borrowing money like this. If oil prices fall, you have no money to pay off the debt.
In 1982 the price of a barrel of oil was $30. By 1986 it had fallen to $15.

C

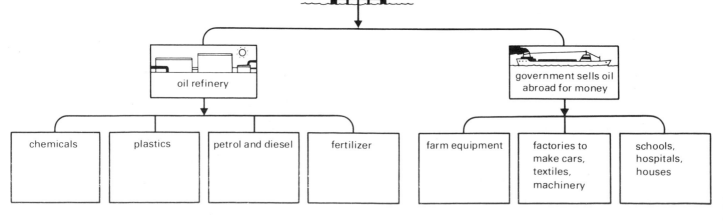

oilfield

oil refinery

government sells oil abroad for money

| chemicals | plastics | petrol and diesel | fertilizer | | farm equipment | factories to make cars, textiles, machinery | schools, hospitals, houses |

Things to do

1 Copy map **B**, which shows the oilfields in Mexico.

2 There are five statements about Mexico below. Write down each statement, and after it write down a fact or figure from the opposite page to prove that the statement is true.
a The Mexican people have a new resource.
b It is difficult to farm in Mexico.
c Mexico City is a large city.
d Jobs arc hard to find in Mexico City.
e Soon there will be many more people living in Mexico.

3 The people in Mexico must make up their minds how best to use their oil. They have two choices, as cartoon **D** shows.

D Where Mexico's oil will go

Use up the oil as quickly as possible

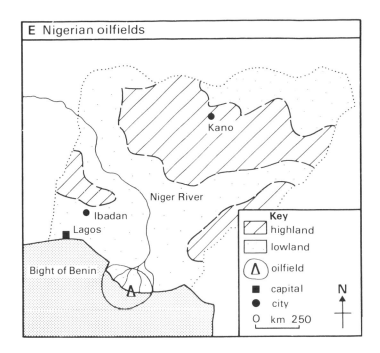

Use the oil slowly over many years

Which of these two ways do you think would be the most sensible? Think about these questions:
How fast do you want to become rich?
How long do you want to stay rich?
What will you do when the oil is all gone?

4 Oil has also been found in Nigeria, a country in Africa. Map **E** shows the Nigerian oilfields. Use map **B**, map **E** and table **F** to help you compare Nigeria and Mexico.

Write four paragraphs to say how the two countries are either the same or different in each of the following ways:
where the oilfields are (Are they on land, sea, inland or along the coast?)
where the capital city is (Is it in the centre or by the coast?)
how many people there are (How many millions?)
how rich the people are (what is the average value of goods produced per person, per year?)
what most of the people do for a living (Do they work on farms or do other types of work?)

E Nigerian oilfields

Kano

Niger River

Ibadan

Lagos

Bight of Benin

Key
〰️ highland
▢ lowland
Λ oilfield
■ capital
● city
0 km 250

N

F	Mexico	Nigeria	UK
Number of people	70 million	80 million	56 million
Number of people in the largest city	15 million	1½ million	7 million
% of people who work in farming	39%	67%	3%
Average value of goods produced per person per year	$2250	$870	$9110

Advanced work

5 If the Nigerian people spend the money they get for the sale of oil wisely, the country could soon change (read question 4 again).
a Draw the map of Nigeria as it is now (map **E**).
b Mark in where you think five industries using oil as a raw material could be built.
c Draw in any new roads, railways, and pipelines which might be needed.

Summary
The discovery of oil has made the future look brighter for Mexico. Countries have to choose how best to use their resources.

Unit 4.3
Trees, an infinite resource

Trees are one of the most important resources we have. Even left in forests they are useful in all sorts of ways. For one thing, they protect the soil. Where forests are cut down the soil is often washed away by rain.

Just one tree is a natural home for thousands of animals. Cutting it down reduces their chances of finding a home and so reduces the richness of wildlife. The great forests also act as the 'lungs' of the world because they put oxygen back into the air. In Western Europe, forests are often visited by tourists wanting to relax in quiet surroundings. Nature trails and clearings are organized to make these visits more interesting.

When trees are cut down, they can be used for firewood. Many people in Europe, Africa and Asia use no other fuel for heating their homes. In Europe and North America the main uses of wood are as timber for building and **wood pulp** for making paper and chemicals (see cartoon **A**).

A The use of trees

Map **B** shows the great hardwood and softwood forests of the world. The wood of most trees growing in warm and wet places is hardwood. Your thumbnail would not scratch their surface. This type of wood is used to make furniture and ornaments. Here are some of the types of hardwood: beech, oak, mahogany, teak.

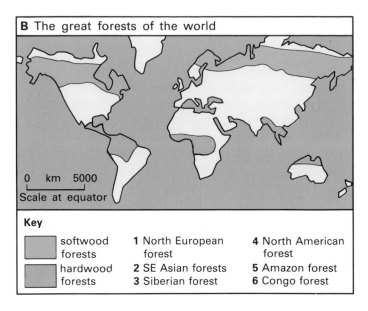

B The great forests of the world

Key

softwood forests	**1** North European forest	**4** North American forest
hardwood forests	**2** SE Asian forests	**5** Amazon forest
	3 Siberian forest	**6** Congo forest

0 km 5000
Scale at equator

Softwoods usually grow in cold lands. Their wood is easily scratched with a thumbnail. Unless it is nice to look at, like pine, it is not used for furniture. Softwoods bear cones. Here are some softwoods: fir, pine, spruce, cedar.

Some resources are finite, and some are infinite. Trees can be an infinite resource.

A **finite** resource is one which runs out. A resource like oil which is burned away and cannot be replaced is bound to run out sometime. So it is called a finite resource.

An **infinite** resource is one which does not run out, but goes on for ever. A forest can go on for ever, if a new tree is planted for every tree that dies or is cut down. If it is looked after carefully, a forest can be an infinite resource (see sketch **C**).

C A finite forest

tree being felled

An infinite forest

tree being planted to replace the felled one

Individual trees are cut down but the forest stays the same.

1 Copy map **B**, which shows the great forests of the world.

2 Answer each of these questions with one sentence. The sketches in **D** may help.
   a What use is one tree when it is just standing there?
   b What use is a forest when it is just standing there?
   c Is oak a hardwood or a softwood?
   d What is wood called when it is crushed and mixed with water into a thick paste?
   e Does the graph in **E** show a finite or an infinite resource?

D The use of a resource

It protects the soil from heavy rain.

A tree is like a city for wildlife.

Forests protect the soil. People like trees.

'Hearts of oak are our ships...'

Wood pulp mill

E Graph of a resource

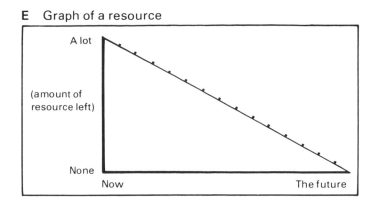

A lot

(amount of resource left)

None

Now   The future

3 Draw a cartoon to show a use for trees which is not shown in **A** or **D**.

4 The great softwood forests are being protected as infinite resources. Unfortunately the great hardwood forests are being cut down rapidly. Use your Atlas to make a list of those countries which may have parts of the great hardwood forests. Write a short protest letter in your exercise book as if you were a conservationist (see note F) writing to the government of one of these countries.

Note F
A **conservationist** is a person who wants to see people living *with* the natural world, not destroying it.

5 Invent a diagram or sketch or cartoon which will show what a finite resource is. Clues **G** may help.

Clues G
What do you buy at a filling station? (Expected to run out in 100 years or so at the present rate)
What powers a nuclear submarine? (100 years running-out time)
What metal is often used as wire for conducting electricity? (50 years running-out time)

Summary
Trees are an important world resource. If a forest is well looked after, it can be an infinite resource.

Uranium, a mineral resource

Any rock which is quarried, mined or drilled into may be called a **mineral** resource. An important mineral resource is uranium ore. It can be melted down to give the metal uranium.

Uranium is important because it is the raw material for nuclear power. Uranium atoms can be made to split apart in a controlled way, giving out a lot of heat energy. One tonne of uranium gives the same heat as 30 000 tonnes of coal (see diagram **A**). Nuclear power, then, is power from uranium atoms.

A Uranium versus coal

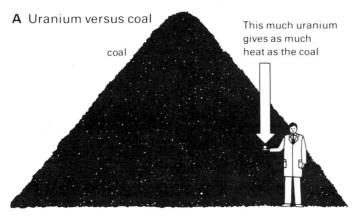

coal

This much uranium gives as much heat as the coal

Some people believe that although nuclear power can be useful, it is too dangerous. Nuclear power stations are quite safe, but the used uranium is radioactive. It gives out harmful rays. The problem is where to store the radioactive uranium.

Nuclear power is also used for military purposes. In 1945 two Japanese cities, Hiroshima and Nagasaki, were destroyed by just two atomic bombs. Most people were killed by the blast. Others, many miles away, died when harmful 'fall-out' from the explosion gave them skin disease.

Uranium is found in the earth by geologists. A **geologist** is a scientist who studies the rocks that make up continents and ocean floors. He studies rocks by going out 'in the field', taking rock samples and drawing maps of rock types in the landscape. He is helped by satellite and aerial photographs and by other geologists' reports. Other detailed work is done in the laboratory. Geologists work for governments and also for mining companies. They find most new mineral resources, though every now and again someone stumbles on a new mineral 'find' by accident. Cartoon **C** shows how this find might be developed. Map **D** on the opposite page shows where uranium is found.

B The work of a geologist

Laboratory work

Field work – going out and looking

Reading what other scientists say

Studying aerial and satellite photographs

C The development of a resource

Discovery

Raising money

Building a quarry or mine

Transporting the resource to where it is sold or used

1 Put the sentence tails to the correct sentence heads.

a Nuclear power is

b One tonne of uranium gives

c The top uranium producer is

d The tenth uranium producer is

e A rock which can be quarried or mined is

a mineral resource.

Namibia.

power from uranium atoms.

the USA.

the power of 30 000 tonnes of coal.

2 Copy map **D** and the first two paragraphs on the opposite page.

E.

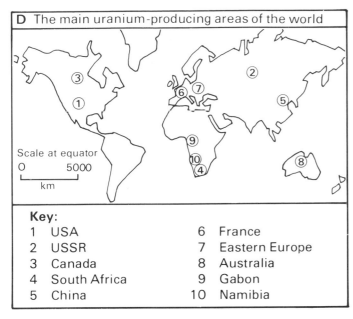

D The main uranium-producing areas of the world

Scale at equator

0   5000

km

Key:

1	USA	6	France
2	USSR	7	Eastern Europe
3	Canada	8	Australia
4	South Africa	9	Gabon
5	China	10	Namibia

3 Copy and complete the part-sketch **F** of White Mine, Rum Jungle, near Darwin in Australia, which is shown in photograph **E**.

4 Write a story to say how you found a rich mineral resource and developed it. Use an Atlas to find out real place names in Australia.

5 Look at the diagram of a nuclear power station (**G**). Make a list of the things that happen. Start with the raw uranium and end with used uranium and electricity.

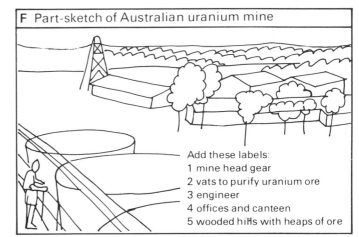

F Part-sketch of Australian uranium mine

Add these labels:
1 mine head gear
2 vats to purify uranium ore
3 engineer
4 offices and canteen
5 wooded hills with heaps of ore

Summary

A mineral resource comes from the rocks of the earth's crust. Uranium is an important mineral resource. It is used to create nuclear power.

G How a nuclear power station works – much simplified

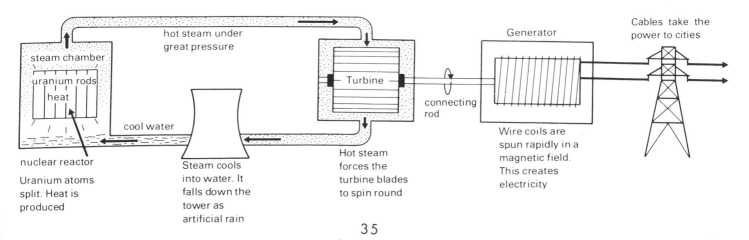

steam chamber

uranium rods

heat

nuclear reactor
Uranium atoms split. Heat is produced

cool water

hot steam under great pressure

Steam cools into water. It falls down the tower as artificial rain

Hot steam forces the turbine blades to spin round

Turbine

connecting rod

Generator

Wire coils are spun rapidly in a magnetic field. This creates electricity

Cables take the power to cities

Industry/The Ruhr

When people are planning to build a factory, they try to choose the best possible site. Different places have different advantages. One place may have **raw materials** to make things out of. Another may have power to drive machinery, or may be easy to get to.

It is unusual to find a place which has all three advantages: raw materials, power and easy transport. The Ruhr area in Germany had all three.

About 200 years ago, the Ruhr area had nothing but farms, villages, and market towns (sketch **A**). Then coal and **iron ore** were found. Before long, there were coal mines and iron foundries, where iron is made from ore. Making **steel** from the iron came next. Then more factories were built to make things out of iron and steel. The workers made trains, bridges, machinery and all kinds of other metal goods.

The River Rhine became the busiest waterway in Europe as barges brought **goods** to and from the Ruhr factories. Soon there was not much farmland left. Instead there were cities, factories, houses and railways (sketch **B**).

Mines and iron factories were just the start. Factory owners became rich and used their **profits** for investment in new factories. More and more people came to the Ruhr, so food, clothes and gadgets of all sorts had to be provided. Shops, hospitals and other **services** were also needed. The Ruhr area is now one of the most built-up parts of Europe. About 10 million people live there (map **C**).

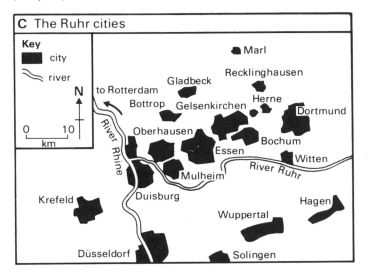

C The Ruhr cities

Key
- city
- river

N

to Rotterdam

0 10 km

Marl
Recklinghausen
Gladbeck
Herne
Bottrop
Gelsenkirchen
Dortmund
Oberhausen
Bochum
Essen
Witten
River Ruhr
Mulheim
River Rhine
Krefeld
Duisburg
Hagen
Wuppertal
Düsseldorf
Solingen

The iron ore is all gone now. Many of the coal mines have been closed down. New industries are taking their place. There are factories making cars, chemicals and electrical goods. These new factories mean that the Ruhr area is still one of the most important **manufacturing** regions in the world.

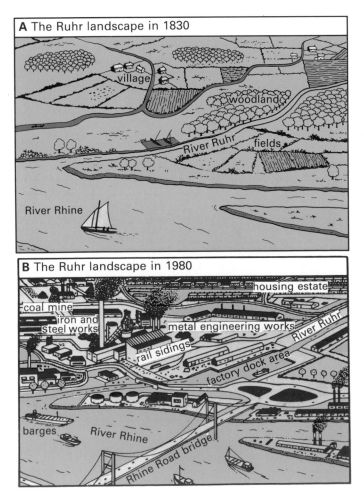

A The Ruhr landscape in 1830

village
woodland
River Ruhr
fields
River Rhine

B The Ruhr landscape in 1980

housing estate
coal mine
iron and steel works
metal engineering works
River Ruhr
rail sidings
factory dock area
barges
River Rhine
Rhine Road bridge

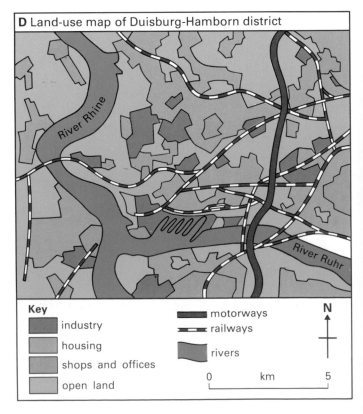

D Land-use map of Duisburg-Hamborn district

River Rhine
River Ruhr

Key
- industry
- housing
- shops and offices
- open land
- motorways
- railways
- rivers

N

0 km 5

1 Draw up a chart to show how jobs in the Ruhr area have changed. Do this by sorting out the following jobs into the right order, putting the earliest first:

E

cloth worker farmer steel worker iron worker miner car worker

2 Answer the following questions with a sentence for each:
  a In which country is the Ruhr area?
  b Name the river which flows past the Ruhr.
  c What two raw materials brought industry to the area?
  d Name two things made of metal made in the Ruhr factories.
  e Give an example of one of the Ruhr area's new industries.

3 The scenery in the Ruhr area has changed over the years. Sketches **A** and **B** illustrate some of the changes.
  Describe as many ways as you can see that the area has changed. Use these sub-headings for your answer:
      settlement (where people live)
      industry (factories)
      farming (growing crops and rearing animals)
      population (number of people)
      transport (travel)
      pollution (dirty waste from a factory)

4 Study the land-use map of the Duisburg Hamborn district (**D**).
  a Trace the frame of the map on tracing paper.
  b On your tracing paper, mark in twenty small crosses scattered anywhere on the map.
  c Lay the tracing paper over the land-use map and count the number of crosses which fall on each type of land-use. Multiply that number by five. This gives you the percentage of your crosses which fell on that type of land-use. Fill in the chart at the top of the page. (The first line has been done as an example, but you should fill in your own numbers.)

Land-use	Crosses	x 5	%
Industry	3 (*example*)	3 x 5 = 15	15%
Housing			
Shops/offices			
Open land			
Railway/road/river			

  d Draw a bar graph to show the percentage.
  e Write a conclusion to say what your graph shows. Say what takes up most of the land, and what is not so important.
  Compare your answers with others in the class who put crosses in different places.

Advanced work

5 Draw an industrial map of the Ruhr area in the following way.
  First copy map **C**. Name the cities but do not shade them in.
  Second, invent symbols for each of the industries listed on the chart below. A small ship drawing, for example, could show where ships are made.
  Third, plot your symbols near to the cities where each thing is made.
  Lastly, don't forget to draw a key to explain your symbols.

Industry in the Ruhr	Iron and steel	Coal	Chemicals	Textiles	Engineering	Ships	Cars
Duisburg	✓		✓		✓	✓	
Essen	✓				✓		
Krefeld				✓			
Bochum	✓				✓		✓
Oberhausen		✓			✓		
Wuppertal				✓			
Gelsenkirchen		✓			✓		
Dortmund	✓				✓		
Recklinghausen		✓					
Düsseldorf	✓		✓		✓		
Solingen	✓				✓		
Bottrop		✓					
Marl			✓				

Summary
A large successful industrial area will attract new industries. The new factories may keep the area rich even when the old industries have faded away. The Ruhr is still Europe's largest manufacturing area, even though many of its old industries have gone.

Unit 5.2
Siberia

The Trans-Siberian Railway crosses the USSR (Union of Soviet Socialist Republics) from Moscow in the west to Vladivostok in the east. It takes eight days to travel the 8000 km.

A The industrial 'beads' on the Trans-Siberian Railway

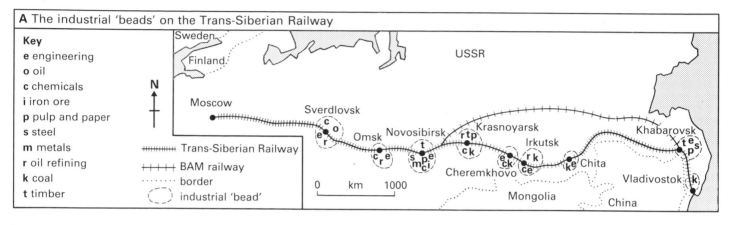

Key
e engineering
o oil
c chemicals
i iron ore
p pulp and paper
s steel
m metals
r oil refining
k coal
t timber

〰〰〰 Trans-Siberian Railway
ⲡⲡⲡ BAM railway
········· border
⌇‾⌇ industrial 'bead'

The train runs through an empty wilderness of flat grassland plains and rolling forests. The winters are bitterly cold, with temperatures well below freezing point.

Siberia is not a useless wasteland. The area has many valuable raw materials. Some, such as wood, are easy to get at. The minerals in the rocks are much harder to find and dig up.

The Russian people have begun to make use of these raw materials. There are quarries and mines to get at coal, iron ore, copper and other minerals. Oil and gas have been found, and rivers are being used to make electricity.

Like beads on a string, towns and cities are being built along the Trans-Siberian Railway route (map **A**). Each 'bead' has factories which turn the local (nearby) raw materials into finished goods. Diagram **B** shows a single 'bead'.

B A Siberian industrial complex (a bead on the string)

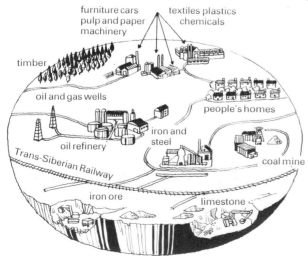

Iron and steel are made by smelting the raw iron ore. Coal gives the heat to do this. The metal is then used in engineering factories to make goods and machinery of all kinds. Oil is refined to give petrol, textiles and plastics.

The Russian government wants people to go and live in Siberia, in spite of the harsh climate. (See photograph **D** on the opposite page.) The area does have some good points, but many people find life too hard and move away again. Cartoon **C** gives some people's views of Siberia.

A small number do not have any choice. They have been sent to Siberia to work in Russian prison camps.

1 Study photograph **D**. Draw a sketch of the photograph and label it, or write a description of the scene.

2 Write out and complete each of the following sentences. The missing words come from the list below.
   a Siberia is part of the ....
   b The Trans-Siberian Railway links Moscow, the capital city, to .... in the east.
   c Siberia has many .... which can be made into different things.
   d Along the route of the railway, there are .... 'beads' of industry.
   e On average, each bead is about .... km from the next.

> nine, 1000, USSR, Vladivostok, raw materials

3 Study **A** showing the industrial beads along the Trans-Siberian railway.
   Draw the two charts, below and on the right. Complete the charts by putting ticks in the columns. Sverdlovsk has been completed for you.

Industrial bead	The resources of Siberia			
	oil	iron ore	coal	timber
Sverdlovsk	✓			
Omsk				
Novosibirsk				
Krasnoyarsk				
Cheremkhovo				
Irkutsk				
Chita				
Khabarovsk				
Vladivostok				

The industries of Siberia					
	pulp and paper	steel	metals	chemicals	engineering	oil refining
Sverdlovsk				✓	✓	✓
Omsk						
Novosibirsk						
Krasnoyarsk						
Cheremkhovo						
Irkutsk						
Chita						
Khabarovsk						
Vladivostok						

4 Imagine you are one of the people shown in cartoon **C** on the opposite page. Write a letter to a friend who lives in Moscow. Try to describe where you live, the work you do, and whether you are enjoying living in Siberia. Begin 'Dear Ivan'.

Advanced work
5 For each of the following raw materials, make a list of at least three things which can be made from it:
   oil, iron ore, timber

Summary
Siberia is an area in the USSR which is rich in raw materials. Although living conditions are difficult, mines, factories and cities are all being built in Siberia.

Unit 5.3
Japan Incorporated

A

Look at the cartoon family above.

Dad drives a car made in Japan. Mum's washing machine was also made in Japan. Their watches, the children's toys, the TV set, record player and camera were all made in Japan. This is a British family, so there must be some very good reasons why they own so many things made in Japan.

After the Second World War, houses, factories, roads and railways all had to be rebuilt in Japan. At first the United States helped by giving money. Factories began to make cheap copies of everyday things which people need. These things were exported all over the world (see map **B**). The Japanese companies made profits and began to find new things to make. Now Japanese workers make everything from pocket calculators to giant oil tankers.

Japan has few resources except the skill of its people. It has to import huge amounts of crude oil from the Middle East, for instance.

C Japanese industry		
Japan makes more of these than any other country:	Japan is second in making:	Japan is third in making:

radio sets
TV sets
ships
	cars
	commercial
	   vehicles
	   (e.g. lorries)
	rubber
	man-made fibres
	   (e.g. terylene)
	plastics
	synthetic rubber
	   (made from oil)
	copper
		pulp (from
		   wood)
		aluminium
		steel

Japanese factories are well organized to produce high quality goods at low cost. Each worker becomes expert at doing just one job on a **production line**. The different parts of the product are put together bit by bit as it moves past on a conveyor belt. Inspectors make sure that faults are spotted and put right.

By this method goods can be produced quickly and cheaply in large numbers. This is called mass production. It is used in modern factories all over the world.

Japanese people are very skilful and take pride in what they make. The workers are loyal to their company. They do not move from company to company like many workers in Britain. Strikes are very unusual, so orders can be delivered on time.

In return the workers are well paid, but that is not all. Big companies also provide houses and entertainment. The managers make sure that the workers understand how the factory is being run. This makes everyone feel part of a team, and nobody wants to let down their team-mates.

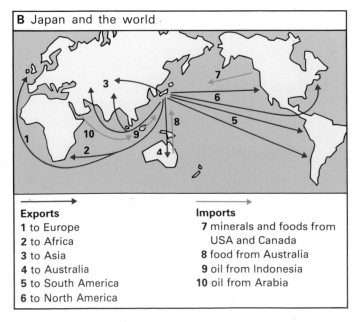

B Japan and the world

Exports
1 to Europe
2 to Africa
3 to Asia
4 to Australia
5 to South America
6 to North America

Imports
7 minerals and foods from USA and Canada
8 food from Australia
9 oil from Indonesia
10 oil from Arabia

Things to do

1 Pair up the 'tails' with the correct 'heads' in the following sentences:

a Mass production	bring in money from other countries.
b American money	means goods can be produced cheaply in large numbers.
c Skilled workers	helped pay for Japan to rebuild its industry.
d Pride in the job	can make accurate and reliable goods.
e Exports	keeps everyone happy and working hard.

2 Pick out at least ten things which are made in Japan from the text and diagrams on the opposite page. Illustrate these things and label your drawing 'Made in Japan'.

3 The two TV sets in sketch **D** are on sale in your local shopping centre. Each has some good points and some bad points.
   a Make a list of the things which you think are most important when you are deciding which to buy. Put the most important thing first.
   b Which one would you buy?

D

E

Advanced work

5 Imagine you are the manager of a big company which makes transistor radios. Your company has made a large profit during the year. You must decide what to do with the money. Here are your choices:
   A Give all the workers a pay rise.
   B Spend the money on research for a new product.
   C Put new machines in your factory which will do the work more quickly and cheaply.
   D Advertise your products more and send more salesmen to other countries.
   E Build a new sports hall and canteen for the workers in your factory.

   a Either on your own or in a group, write down the advantages and disadvantages of each of these choices.
   b Write a report to say how you would spend the money.

4 Explain how you think each of the following helps a Japanese company to be successful:
   – skilful workers
   – pride in the job
   – loyalty
   – team work

Summary
Japan has become one of the world's most important manufacturing countries. Japan has become successful because the people have worked hard and because they have been well organized. Companies have become rich by selling goods all over the world.

Unit 5.4

Space-age industry

*'That's one small step for a man,
one giant leap for mankind.'*

Neil Armstrong the American astronaut spoke these words as he stepped onto the moon from the Apollo XI lunar module on 21 July 1969. The world was amazed at the bravery of the astronauts, and the skill of the scientists and engineers working with NASA (National Aeronautics and Space Administration). People who work on aircraft and space vehicles are all part of the **aerospace industry**. The aerospace industry is one of the most modern engineering industries in the world.

The US aerospace industry has centres in many places in the United States (see map **A**). Each place has a special job. Some of the places have been specially built by NASA. The others are contract firms which have to be asked to do particular jobs, such as making the rocket engines.

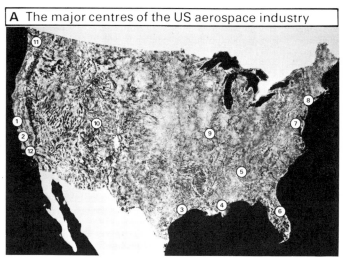

A The major centres of the US aerospace industry

Key

City	State	NASA centre or contract firm
NASA centres		
1 San Francisco	California	Ames Research Centre
2 Vandenberg Air Base	California	USAF Launch Centre
3 Houston	Texas	Johnson Space Centre
4 New Orleans	Mississippi	National Space Technology Labs
5 Huntsville	Alabama	Marshall Space Flight Centre
6 Cape Canaveral	Florida	Kennedy Space Centre
7 Washington	Maryland	Goddard Space Flight Centre
Contractors		
8 New York	New York	Grumman, CEC, RCA, Fairchild
9 St Louis	Missouri	McDonnell-Douglas
10 Denver	Colorado	Martin-Marietta
11 Seattle	Washington	Boeing
12 Los Angeles	California	Rockwell International, Lockheed, Jet Propulsion Lab

Cape Canaveral is the rocket launch base. Building and testing the rockets takes place in California over 1000 km away. People in many other places also help.

Information is passed from place to place by computer links. Parts for the satellites are usually small and can be flown from place to place. The large rocket parts are flown in specially designed aircraft with a very wide body.

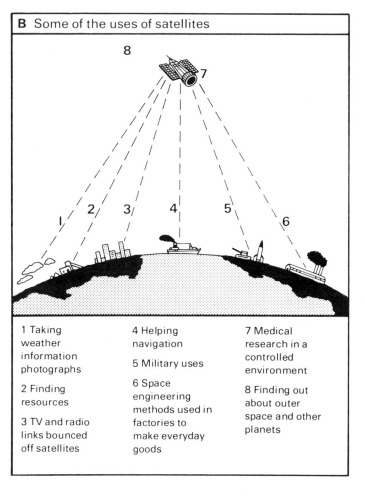

B Some of the uses of satellites

1 Taking weather information photographs

2 Finding resources

3 TV and radio links bounced off satellites

4 Helping navigation

5 Military uses

6 Space engineering methods used in factories to make everyday goods

7 Medical research in a controlled environment

8 Finding out about outer space and other planets

The moon landing was just one sign of what the aerospace industry can do. NASA also sends up satellites for other reasons (see sketch **B**).

Photograph **A** was taken by a satellite called Landsat. This satellite was put up to study the earth's resources. A photograph like this can be made very much larger so that more details can be seen.

Satellite photographs can help find places where there are raw materials, or they can give information about crops and soil. Countries which do not have an aerospace industry can buy photographs from NASA.

The aerospace industry also makes rockets and satellites for military uses. 'Spy' satellites watch what is going on in other countries. The US and the USSR have missiles always ready in case they go to war.

1 Copy map **A** from the opposite page. Mark the names of the cities and the names of the NASA centres and contract firms.

2 Answer each of the following questions with a sentence:
   a What does the term 'aerospace industry' mean?
   b How many NASA centres are there?
   c What is a contract firm?
   d How do the different places keep in touch with each other?
   e Why do you think that aircraft are used to carry satellite parts from place to place?

3 Study the information on map **C**. Explain why you think each of the following chose their location:
      Kennedy Space Centre
      Grumman
      Rockwell International
      Martin-Marietta

4 **a** Describe how each of the people in sketch **D** could be helped by satellites.
   b Make your own list of people who might benefit from satellites, and say how satellites could help them.

Advanced work

5 Putting men on the moon took millions of dollars. The money could have been spent on helping poor or hungry or homeless people on earth. On the other hand the USA spends less than 2% of its annual budget on space programmes. This is less than European women spend on lipstick.
   What are your views on spending money on space research?

C About the space centres

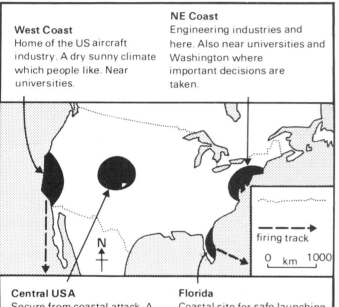

West Coast
Home of the US aircraft industry. A dry sunny climate which people like. Near universities.

NE Coast
Engineering industries and here. Also near universities and Washington where important decisions are taken.

firing track

0 km 1000

N

Central USA
Secure from coastal attack. A nice place to live in – mountain scenery and pleasant climate.

Florida
Coastal site for safe launching over sea. Sunny weather but occasional hurricanes.

Summary
The United States aerospace industry is a very advanced industry which needs scientists and skilled engineers. It works from many different places around the United States. The industry is helping to solve many of man's everyday problems.

D

Unit 6.1
Cities/Moscow, a capital city

Think of Russia and you think of Moscow. Moscow is the Russian city we usually see on TV reports. On May day every year there is a parade through Red Square in Moscow. Heads of governments hold important meetings in Moscow. The 1980 Olympic Games were held there.

Moscow used to be in the very centre of Russia. The Russian people lived in a broad and fertile plain, and all roads led to Moscow. The word 'location' is used to say where a place is.

Now Russia is only part of a much larger country called the USSR. Moscow is not in the centre of the USSR, but has become its **capital city**. A capital city is where the government has its main offices. In Moscow, the government offices are in the Kremlin, the oldest part of the city. Political decisions taken here can affect the whole world.

Moscow is the largest city in the USSR. About 9 million people live here, out of 281 million in the whole USSR. Moscow covers 900 km², which makes it one of the largest cities in the world.

Moscow is a centre for communications. This means that road, rail and air routes from all over the USSR meet here. It is the centre for culture and entertainment, as well as for government. The largest universities, libraries and museums are in Moscow. Television, radio and newspapers all have their headquarters in the city. It is the most important tourist centre in the USSR, and the most important industrial centre. These things are some of the functions of Moscow. It also has the functions of an ordinary city. It provides housing, jobs and services for the people who live there.

For the average Muscovite (a person who lives in Moscow), home is a two-roomed flat in a five-storey block. The government says that every worker has the right to somewhere to live, not too far from where he or she works. Thousands of housing blocks which all look the same have been built. The city may look dull and boring over large areas, but at least everyone has a home.

To get to the city centre, there is an underground railway service (Metro). Muscovites do not own as many cars as people in British or American cities. Cars in the USSR are expensive and difficult to buy.

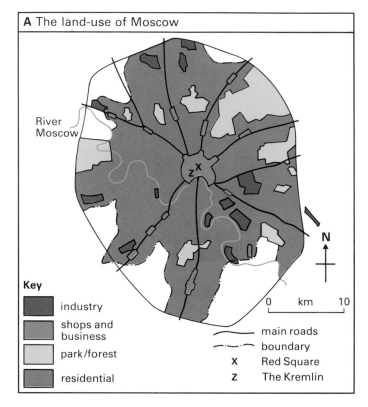

A The land-use of Moscow

River Moscow

N

0   km   10

Key

industry

shops and business

park/forest

residential

main roads

boundary

X   Red Square

Z   The Kremlin

B

44

1 a Draw cross-section **C** (below).
 b Label in the following in the numbered boxes:
   1 The Kremlin   5 Blocks of flats
   2 River Moscow   6 Park and forest
   3 Business centre
   4 Industry

C Cross-section through Moscow

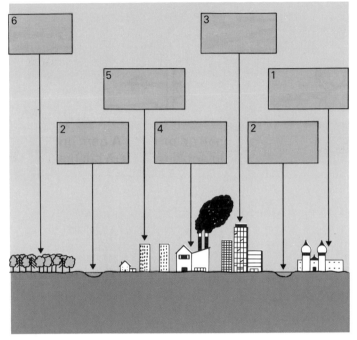

Advanced work

5 a For each of the following functions of
Moscow, make a list of the buildings which
are needed. Some of the buildings are given
below, but you should add more.

Example: **Function**   **Building**
          education   university

Functions:
administration
entertainment
culture
education
communications
industry
housing

> university . . . offices . . . factories . . .
> warehouses . . . railway station . . . airport.

b Write out the list of functions again.
Make a list of the buildings for each function
which are found in your own city, town or
village. Be careful not to confuse the
local district you live in with the whole town
or city.
c Write a paragraph to compare what there
is in Moscow with what there is where you
live.

2 Pair up the 'tails' to the correct 'heads' in the
following sentences:
   a Moscow is in the centre of   in small flats.
   b The government of the USSR   out of 281 million people in the USSR.
   c The Kremlin is in   a broad fertile plain.
   d There are 9 million Muscovites   is based in Moscow.
   e Many of the people in Moscow live   the centre of old Moscow.

3 Here are some other types of location for capital cities. They are jumbled up. Use an Atlas
to fit each of these capitals to the correct description:

Capital	Description
Madrid (Spain)	the biggest port in the country
Rome (Italy)	where one of the world's longest rivers splits up before it reaches the sea
Lagos (Nigeria)	built recently inland, away from the country's other cities on the South Atlantic coast
Brasilia (Brazil)	an ancient city, which stands half-way up a long narrow country
Cairo (Egypt)	right in the centre of the country, where everyone can get to it easily

4 Imagine what it would be like to live in one
of the blocks of flats shown in photograph **B**.
   a Give reasons why you would, or would not,
   like to live in one of these blocks.
   b Why do you think so many buildings of the
   same type have been built?

Summary
The capital city of a country is often the
largest city. It is the centre for many important
activities. Some capital cities are located in
the centre of the country, but for reasons of
history this is not always true.

Unit 6.2
The port of Vancouver

In 1792, Captain George Vancouver of the Royal Navy ordered his crew to drop anchor in a sheltered **inlet** off what is now part of Canada. He was exploring the West coast of North America for the British Admiralty. At that time the area was uncharted and not much was known about the place. Red Indian tribes were the only people who lived there.

One hundred years later, there was a growing city on the site. (A **site** is the exact spot where something is built.) The city was named Vancouver and it is now the busiest port on North America's Pacific coast. About a million people live in Vancouver. Many of them work in offices, warehouses and factories which have been built because of the docks. Map **A** shows some of Vancouver's import and export trade.

The photograph and small map (**B**) show Burrard Inlet and part of Vancouver's docks. The inlet is sheltered from the waves of the Pacific Ocean by Vancouver Island. The water is deep enough for most large cargo ships, and it does not become blocked by ice in winter. Flat land by the inlet has made it easy to build **quays** and all the other buildings needed in a port city.

The Rocky Mountain range rises up behind the city. These mountains could have been a barrier between Vancouver and the rest of Canada. Luckily, rivers have cut valleys through the mountains. This has made it easy to build roads and railways.

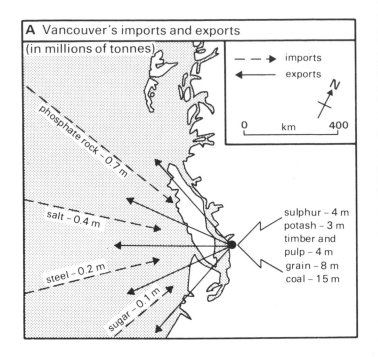

A Vancouver's imports and exports (in millions of tonnes)

imports
exports

0 km 400

phosphate rock – 0.7 m
salt – 0.4 m
steel – 0.2 m
sugar – 0.1 m

sulphur – 4 m
potash – 3 m
timber and pulp – 4 m
grain – 8 m
coal – 15 m

The word **accessible** is used to say how easy or hard it is to get to a place. Road and railway links through the mountains have made Vancouver a very accessible place.

Nature has helped Vancouver become a port. Ships are safe in the natural **harbour** of Burrard Inlet. Goods can be brought to and from the city fairly easily.

Nature has also given the people who live in Vancouver a very attractive landscape to look at and visit. Sea, islands and snowcapped mountains are all within 5 km of the city.

B

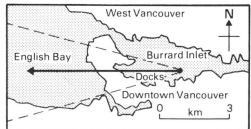

Direction of photograph

West Vancouver
English Bay
Burrard Inlet
Docks
Downtown Vancouver

0 km 3

Key

1 Downtown Vancouver
2 English Bay
3 Canadian Pacific Railway terminus
4 Canadian National Railway terminus
5 Coal Harbour
6 Deadman's Island
7 Stanley Park
8 40-tonne crane
9 30-tonne crane
10 Burrard Inlet
11 15 m minimum depth in channel
12 Centennial Pier
13 container depot
14 rail sidings
15 mobile cranes
16 tug boat
17 container ship
18 Ballantyne Pier
19 general cargo

1 a Draw the cross-section **C** through Vancouver (below).
b Label the following in the right place:

1 Pacific Ocean	5 Burrard Inlet
2 Vancouver Island	6 Vancouver City
3 Georgia Strait	7 Fraser Valley
4 Stanley Park peninsula	8 Rocky Mountains

C Cross-section through Vancouver

Horizontal scale

0   km   50

west   east

2 Study photograph **B** on the opposite page. Answer each of the following questions with a sentence.
  a What kind of ships are most of the ones you can see in the photograph?
  b How can you tell that the water in the inlet is deep?
  c Why do some of the ships not need the cranes on the quay?
  d What other form of transport ends at the docks?
  e What would tell you that Vancouver is a busy port?

3 The map outline **D** (below) shows the area seen in photograph **B** on the opposite page. Use the small map (inset) to work out directions.
  a Make a copy of the map outline.
  b Draw and label in on your map all the features shown on the photograph.
  c Explain how the photograph shows that Vancouver is a good natural harbour.

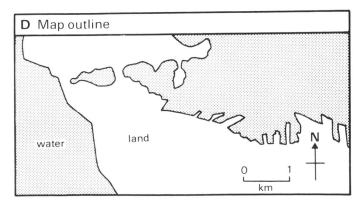

D Map outline

water   land   N

0   1
km

4 The figures on map **A** show the amount and types of export and import goods through Vancouver's port.
  Complete the following paragraph with the right words using information from map **A**.

Vancouver is mainly a port where goods are imported/exported. The main item exported is coal/pulp. Most of the exports are raw materials/manufactured goods. The farm crop which is exported is raw sugar/grain. The main import item is phosphate/salt.

Advanced work

5 Maps **E** and **F** (below) show two of the world's most important port cities.
  a Draw each of the maps.
  b Using the latitude and longitude lines, find the name of each port in an Atlas.
  c Label in the names of all items listed in the key for these maps by looking up each port in an Atlas.

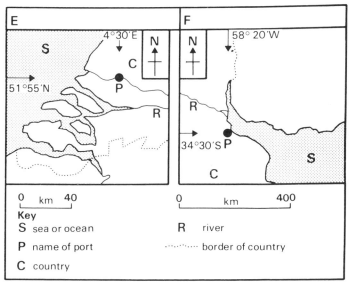

E   4°30'E   N   S   C   51°55'N   P   R

F   58°20'W   N   R   34°30'S   P   S   C

0   km   40   0   km   400

Key

S	sea or ocean	R	river
P	name of port	·········	border of country
C	country		

Summary
Vancouver is a very important port. It has the best natural harbour on Canada's Pacific coast. Many routeways from inland lead to Vancouver. The port has been made by building quays and warehouses.

Lima, the 'do-it-yourself' city

The house shown in sketch **A** is up for sale in Lima, the capital city of Peru in South America. The house was built by a young man and his family. They came to Lima from his father's farm in a poor country area.

At first they rented a room near the city centre. The young man could not find a job, and soon they had no money left.

Next they moved to a patch of land which they heard was being taken over by other poor people. There was a fight with the police, but there were so many **squatters** that it was impossible to make them leave. A squatter is a person who takes over an empty house or land and lives there without owning it or paying rent.

In Lima, these squatter areas are called the **barriadas**, which means a slum neighbourhood. The buildings are put up almost overnight using any scrap materials people can find. As years go by, the barriada houses are improved with brick and concrete. Some people, like our young man, manage to find work and save some money. Now he and his family want to move out to live in a better area.

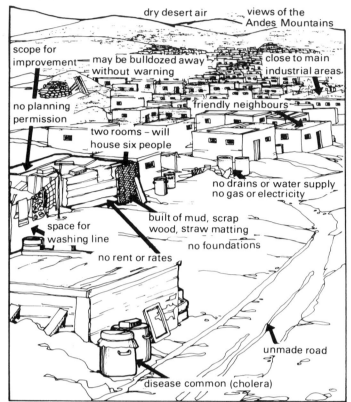

A House for sale in Lima

In Lima, one person in every six lives in the barriadas, sometimes called **'shanty towns'**. Every day, more people arrive from the countryside. This movement of people from the countryside to cities is called 'urbanization'.

By the year 2000, there will be many cities in South America where seven out of ten people live in shanty towns. In a city of 10 million people there will be 7 million miserable lives, unless someone finds an answer to poverty.

C	The population of Lima
Year	Population in millions
1940	0.5
1950	1.0
1960	1.5
1970	2.5
1980	4.5
1990	8.0 est.

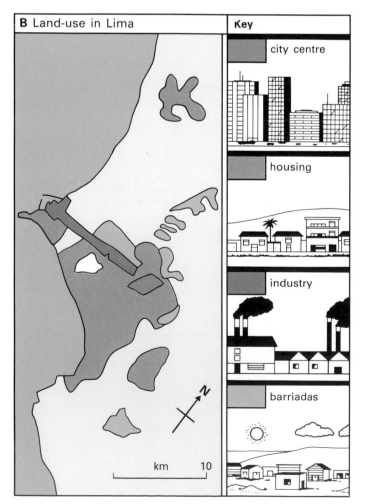

B Land-use in Lima

Key: city centre, housing, industry, barriadas

km 10

1 a Make a copy of sketch **A** of the barriada house in Lima.

b Draw a chart to sum up the good and bad points about this house:

Good points	Bad points
cheap to build	no water supply

2 The figures in table **C** show how more people have come to live in Lima over the years.

a Draw the graph outline **E**.

b Write the title 'The Growth of Lima'

c Plot the population figures on the graph. The first two have been done for you.

d Join up the dots you have marked on the graph with a curved line.

e Now look at the three graphs (**F**). See which one your graph most looks like. Write the description that best suits your graph.

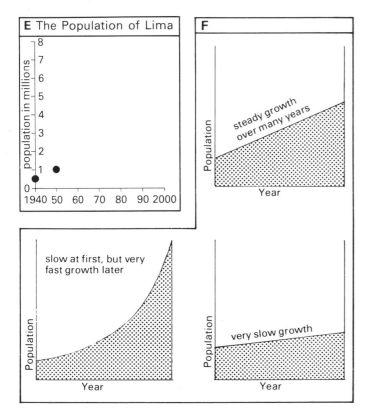

E The Population of Lima

F

steady growth over many years

slow at first, but very fast growth later

very slow growth

3 a Make a copy of map **G**, which shows the ten fastest-growing cities in the world.

b Write out the two statements from the list below which are correct:
The ten fastest-growing cities are
– on the coast
– in one continent
– in poor countries
– in rich countries
– spread over Africa, South America and Asia

c Use an Atlas to find out the name of the country each of these cities is in.

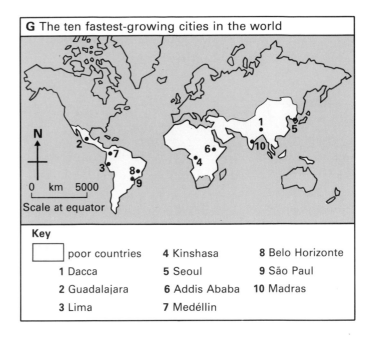

G The ten fastest-growing cities in the world

Key

	poor countries	4 Kinshasa	8 Belo Horizonte
1 Dacca		5 Seoul	9 São Paul
2 Guadalajara		6 Addis Ababa	10 Madras
3 Lima		7 Medéllin	

4 As more people move to the cities, extra buildings are needed to meet everyone's needs. Use the following headings to make a list of the different types of buildings which would be needed:

Jobs
Shelter
Health
Food
Education
Transport

Say why you think it is so difficult to provide all these buildings in a poor country.
Think about how much money could be needed to build one new hospital or school, and how long it would take.

Advanced work

5 Study the map of land-use in Lima (map **B**). Describe the way the land is being used. Mention the following:
– the different types of land-use
– the amount of each land-use type
– where the different types of land-use are found

Summary
More and more people are moving to live in cities. This movement is called urbanization. Urbanization is mainly happening in poor countries. Great problems are caused when too many people move at the same time. Many people live in dreadful conditions in cities which have grown too quickly.

Megalopolis, USA

The people shown in cartoon **A** (below) have one thing in common. They all live in the megalopolis of the north-eastern United States.

A Office worker Truck farmer Dock worker Factory worker Shop assistant

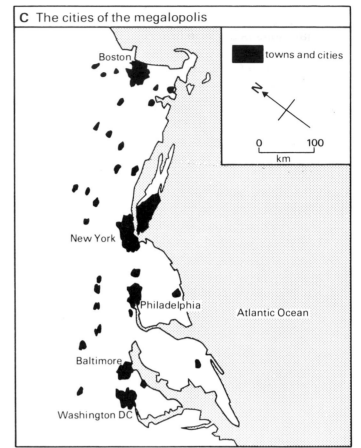

C The cities of the megalopolis

towns and cities

Boston

New York

Philadelphia

Atlantic Ocean

Baltimore

Washington DC

0   100
km

The word 'megalopolis' means 'a great city', but it is far more than just that. It is a huge built-up area stretching 500 km from Boston in the north to the capital city, Washington DC, in the south. There is some farmland and woodland left, but houses, factories and roads are being built on these open spaces. The megalopolis is shown on sketch **B** and map **C**.

The people of the megalopolis have many different ways of life. Some work for the government in Washington. There are dockers in the big port cities of New York and Boston. There are factories in all the cities making electrical goods, clothes, chemicals, steel, and almost everything else. Farmers grow fresh food for the city people and drive it by truck to the city markets. This is why they are sometimes called 'truck farmers'.

The megalopolis is one of the largest areas in the world to be so **densely populated**. About 46 million people live here. Some live in city slum areas where houses and flats are in a dreadful condition. Many of these people are unemployed. The rich people prefer to live around the edge of the cities. These areas are called the suburbs. Suburban houses are large and often detached, standing in their own grounds.

Nobody wants to live in the poor areas in the city centre, so people try to move out to the suburbs or further into the countryside. There are good road and rail links, so travelling back to city offices and factories is not a problem. Travelling like this is called commuting. Some offices and factories are now moving out of the city to find somewhere with more space and clean air.

By the year 2000, there will be an extra 34 million people living in the megalopolis. By then, some of the cities will be completely joined together without a break. There will not be much countryside left. There will be one giant super-city with the one name of Megalopolis.

B Land-use in the megalopolis

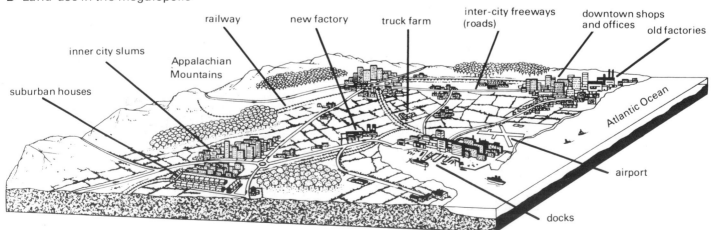

railway   new factory   truck farm   inter-city freeways (roads)   downtown shops and offices   old factories

inner city slums   Appalachian Mountains

suburban houses

Atlantic Ocean

airport

docks

1 Copy map **C**, which shows where the megalopolis is and the different cities in it.

2 Write out the five statements from the list below which are true:
The megalopolis is the capital city of the USA.
The megalopolis is made up of many different cities.
There is some open land left between the cities.
All the cities in the megalopolis are completely joined up.
The people in the megalopolis mostly work on farms.
There is a wide range of office and factory jobs.
Rich people like living in the central parts of the cities.
Most people want to move out of the cities.
Good road and rail links make it easy for people to live in the countryside, but still work in the cities.
The megalopolis is 100 km from north to south.

3 Study the two strip cartoons **D** and **E**.
Each one shows a typical day in the life of a person who lives in the megalopolis. Describe the typical days of these two people.

4 Each of the descriptions below can be replaced by a single word, or two words. Write down the word or words for each one.
A farmer who grows fresh food for the city.
A housing area on the edge of a city.
A house standing alone in its own grounds.
A way of describing a lot of people living close together.
A person who travels to work every day.
An area of poor housing near the city centre.
A road or railway line joining two places.

Advanced work
5 Why do you think that planners in the USA are worried about all the open space being built on in the megalopolis? Use these subheadings for your answer:
　Pollution　　　　　Recreation (play) space
　Growing food　　　Scenery

Summary
Megalopolis is a group of cities in the north-eastern USA which are growing outwards into the countryside. Soon there will be little countryside left. People do not enjoy living in city centre areas and prefer to move out where there is more space and it is more healthy.

D Leaves home in the morning in a big American car. Home is a large house in the country.

Parks the car at the station. Catches a commuter train to the city.

Works in an office block high up.

Comes home. Goes out in the evening for a round of golf.

E Walks from a downtown slum house.

No work – many factories have closed down.

Collects welfare money (American social security). Checks on jobs vacant.

Plays pool in a downtown bar.

People at play/Leisure in the city

'All work and no play makes Jack a dull boy'. This well-known saying is true for most people. **Leisure time** is the time when a person is not working in a school, factory or office. It may be a time for just doing nothing. Usually, it is a time for hobbies, sports or other interests. Sometimes, it is a time for sitting and reading, watching TV, or being entertained by someone else.

There are lots of different ways of spending leisure time in towns and cities (see map **A**). There are parks, adventure playgrounds and youth clubs for children's day-to-day leisure. Most schools also provide something to do as well as the classroom lessons.

B

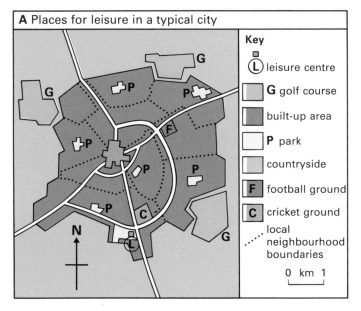

A Places for leisure in a typical city

Key

⌖ Ⓛ leisure centre

G golf course

built-up area

P park

countryside

F football ground

C cricket ground

····· local neighbourhood boundaries

0 km 1

Adults can go to public houses and bingo halls after work. Here they can relax and chat with friends. Some large factories have their own social clubs for their workers.

All these places are near to people's homes. They are used every day, so they must be easy to get to. Places where children can play must also be nearby, because parents prefer their children to play near home.

Leisure time at the weekend may be spent differently. Football supporters travel to see the team they support during the football season. During the summer, cricket fans can go to see their county team.

People who enjoy taking exercise can play golf or some other active sport. Most cities have golf courses on the edge of the city. There is more space there than in the centre, and the land is cheaper to buy.

New leisure centres have been built in many cities. Perhaps your local town has one. Leisure centres are places where you can play the sport you enjoy most, or try out something new. The word **recreation** is used to describe a leisure activity which you take part in, rather than just watching.

Most of the city 'night-life' takes place in the city centre. This is where large cinemas, ice rinks and bowling alleys are found. They are expensive to run, so people have to pay a lot of money to go in. Not many people can afford to go to these places more than once a week.

The city centre is where all the main roads and bus routes meet. It should be the part of the city which the largest number of people can get to easily.

C A street in a city centre

ICE RINK · BOUTIQUE · DISCO CLUB · CINEMA · SUPERMARKET · Dental Surgery · RECORD SHOP · SPORTS CLOTHES · TRAVEL AGENT · BANK · TOBACCO SHOP

Things to do

1 This exercise is called 'giving reasons'. Pair up each of the statements on the left with the reason on the right which explains it best.

Statement

  a Children's playgrounds are not far from each other in most neighbourhoods

  b Golf courses are usually on the edges of cities

  c There are not many theatres in a city

  d Cinemas are mainly in the city centre

  e Even places which a lot of people like, such as bowling alleys, are not visited very often by each person

Reason

because not enough people go to them regularly to support a larger number.

because they are quite expensive, and people have to travel to the city centre to reach them.

because people from all over the city should be able to get there easily.

because parents like children to play near home.

because there is more space there than there is inside the city.

---

2 Draw up a chart to show how you spent your leisure time during the past two days, or longer, and over the last weekend:

Personal leisure chart				
Day	Leisure time	Place for leisure	Kind of activity	Cost
example: Monday	4 p.m. – 5 p.m.	street	chatting	free

3 Study Map **A** and the information on the opposite page. Write out and complete the following sentences. Choose the missing words from the ones below.

  a Most neighbourhoods have their own local ....

  b Most .... are found on the edge of a city.

  c A .... has been built where four main roads meet.

  d Some factories run their own ....

  e There is not much open space near the ....

> leisure centre ... social clubs ... golf courses ... city centre ... park

4 Study sketch **C** on the opposite page.

  a Make a list of all the buildings which have something to do with leisure and recreation.

  b Write a paragraph to say why these buildings are located in the centre.

Advanced work

5 The two diagrams and maps **(D)** show where people come from to go to two different places for leisure.

a Complete the graph outlines **(E)** by counting up how many people have travelled each distance.

b Write a paragraph to say what these graphs show.

D Ten children in a local park

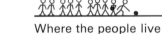

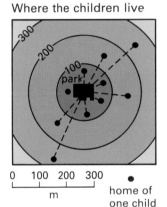

Where the children live

0 100 200 300
m

● home of one child

Ten people in a city centre bowling alley

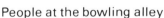

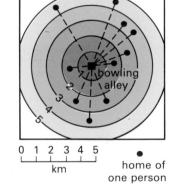

Where the people live

0 1 2 3 4 5
km

● home of one person

E Children at the park

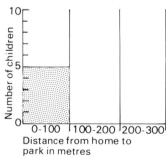

Number of children

0-100 100-200 200-300
Distance from home to park in metres

People at the bowling alley

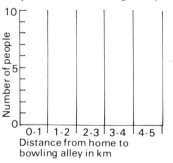

Number of people

0-1 1-2 2-3 3-4 4-5
Distance from home to bowling alley in km

Summary
Leisure takes up a large part of most people's time. In most towns and cities there is a good choice of things to do in free time. There is a pattern to the number, type and location of different leisure places.

Unit 7.2
The great outdoors

People who live in cities have to put up with noise, polluted air and traffic jams. They like to get away from these problems sometimes, and have a change of scene.

Going to the **countryside** is a popular way of using leisure time. More people own cars than in the past, so getting to the countryside is easier. New roads, especially the motorways, have also helped. People can drive much farther in an hour than they could before the motorways were built.

In some countries, National Parks have been opened to give people somewhere to go on their trip out. National Parks are usually in mountain areas where the scenery has not been ruined by too many buildings or people.

Map **A** shows the Rocky Mountain National Park in Colorado, USA. It is visited by 3 million people every year. They come to fish, climb and follow trails on horseback. Some come in winter for skiing and sledging. Most just want to enjoy the clean mountain air and look at the peaks and glaciers. Rare bighorn mountain sheep can sometimes be seen, as well as beavers, bears and golden eagles.

Hunting is not allowed. Park Rangers have to make sure that the wildlife and plants are not destroyed by visitors. Feeding the bears could be bad for the health of both bears and visitors.

It is hard to stop the Park from getting over-crowded. In summer, the roadside picnic sites are soon filled with visitors and their cars. There are not many roads through the mountains, so traffic jams can be expected. Too many visitors can bring the problems of city life into the countryside.

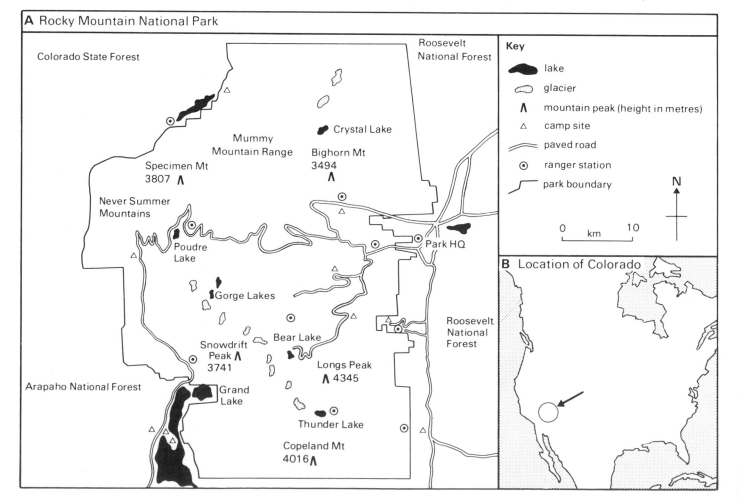

A Rocky Mountain National Park

Colorado State Forest

Roosevelt National Forest

Key
- lake
- glacier
- Λ mountain peak (height in metres)
- △ camp site
- paved road
- ⊙ ranger station
- park boundary

N

0   km   10

Crystal Lake

Mummy Mountain Range

Specimen Mt 3807 Λ

Bighorn Mt 3494 Λ

Never Summer Mountains

Poudre Lake

Park HQ

Gorge Lakes

Bear Lake

Snowdrift Peak Λ 3741

Longs Peak Λ 4345

Arapaho National Forest

Grand Lake

Thunder Lake

Copeland Mt 4016 Λ

Roosevelt National Forest

B Location of Colorado

1 a Make a copy of graph **D**, which shows the number of visitors to US National Parks between 1950 and 1977.
  b Write a sentence to say what the graph shows about the number of visitors in this period.

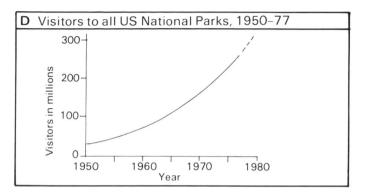

D Visitors to all US National Parks, 1950–77

2 Use the words below to complete the following sentences:
  a The Rocky Mountain National Park is visited by .... million people each year.
  b Visitors are helped by ...., who also have to protect the Park.
  c The peace of the countryside may be spoiled by visitors who mainly come during the ....
  d A change of .... can make a pleasant break.
  e A National Park is a special area for ....

> scene . . . three . . . summer . . . recreation . . . Park Rangers

3 Table **E** shows the percentage of people in the United States who take part in different leisure activities.
  a Make a copy of the table.
  b Put a cross in the columns which you think best describe each activity.
  c Add up the total number of crosses in each column.

d Complete the following sentences using the correct word:
  Most leisure activities take place indoors/ outdoors. People prefer to take part in / watch leisure activities. The town/countryside is the most popular place for leisure activities.

4 Use photograph **C** to draw a poster to advertise the Rocky Mountain National Park. On your poster, try to say what kind of things can be done and seen in the Park.

Advanced work

5 List **F** shows some commonsense 'Do's' and 'Don'ts' which park visitors should obey. Write a short story to describe a day in the life of a Park Ranger coping with people who do not follow the rules. You could use the map on the page opposite to mention names of places and distances travelled.

F Rules of the Park

Do
Use a detailed contour map when walking in the Park
Tell the Ranger if you are going on a difficult rock climb
Stay away from the edge of steep snow slopes
Put all cans and litter in the litter bins (they give the bears a stomach-ache)
Lock your car when away for a walk
Don't
Try to wade in streams in spring when they are deep, cold and fast
Keep food in tents – bears think you are playing hide-and-seek
Light fires unless you have a permit
Go for a two- or three-day hike without leaving a trail route with the Ranger

Summary
People often like to visit the countryside for recreation. New roads have made it easy to drive long distances quickly. Some places become very crowded because they are so popular.

E Leisure activities in the United States	Indoors	Outdoors	Watching	Taking part	Town	Country
Picnicking 72% (example)		✗		✗		✗
Visiting fairs, zoos, parks 72%						
Pleasure driving 69%						
Jogging 68%						
Swimming (in pool) 63%						
Sightseeing 62%						
Going to outdoor sports 61%						
Boating 59%						
Fishing 53%						
Camping 51%						
Snow sports 51%						
Total						

(Figures do not add up to 100% as many people take part in several activities.)

Sunshine holidays

The Greek airliner (photo **A**) is kept busy during the summer months. It helps to bring 4 million tourists to Greece every year. In August at the height of the holiday season, the aircraft are run like buses in a city rush hour. As soon as they land they are refuelled, checked, loaded with passengers and sent off again.

Map **B** shows which countries most of the tourists come from. The thickness of each line depends on the number of tourists from each country. Notice that most of the tourists come from the rich countries of Northern Europe, and from the USA. A map like this is called a **flow map**.

A

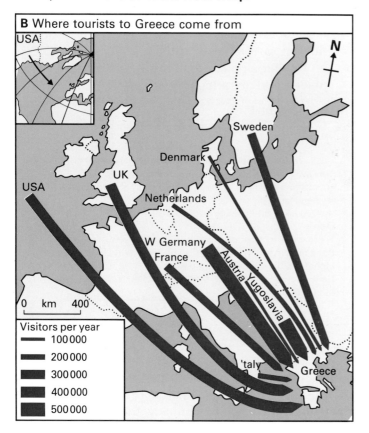

B Where tourists to Greece come from

USA

Sweden
Denmark
UK
USA
Netherlands
W Germany
France
Austria
Yugoslavia
Italy
Greece

0   km   400

Visitors per year
— 100 000
— 200 000
— 300 000
— 400 000
— 500 000

Greece has the kind of summer weather most holidaymakers like. It is usually hot and there is not much chance of rain. It is good weather to get a suntan.

Other countries along the Mediterranean coast have the same kind of summer weather. Spain and Italy are also popular for summer holidays.

Greece has more than just good sunbathing weather. Greece has its own customs, food, costumes and history. All these things add up to make Greece an attractive and interesting place to spend a holiday.

The hotel in drawing **C** (below) is one of many on the Greek island of Corfu. The sketch shows some of the attractions this hotel can offer.

There is a small village nearby where guests can buy souvenirs or just sit outside a taverna (inn) and watch everyone else. There is a special bus service from the hotel to the main town. There are also coach tours to take guests to see the 'sights' all over the island.

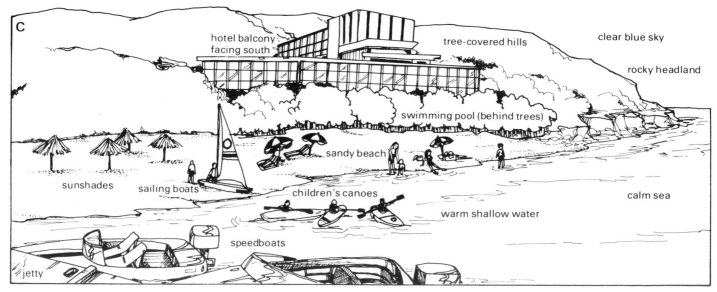

C

hotel balcony facing south
tree-covered hills
clear blue sky
rocky headland
swimming pool (behind trees)
sandy beach
calm sea
sunshades
sailing boats
children's canoes
warm shallow water
speedboats
jetty

1 Arrange the following list in the order you think most important for a successful holiday: sea-bathing, different food, new friends, amusements, historic places, sunshine, different language, scenery.

2 Study map **B** on the page opposite, and table **D** below. They show where people come from to spend a holiday in Greece.
   Write out and complete these sentences:
   a More people come to Greece from . . . . . . than from anywhere else.
   b The same percentage come from the UK to Greece as from . . . .
   c The country on the list which is furthest away from Greece is . . . .
   d Most tourists come to Greece from the continent of . . . .
   e The total percentage who come to Greece from the top ten countries is . . . . %.

D	Tourists in Greece	
	Country of origin	**%**
1	W Germany	14
2	USA	13
3	Yugoslavia	12
4	UK	12
5	France	8
6	Sweden	6
7	Italy	4
8	Austria	3
9	Netherlands	3
10	Denmark	3
	Other countries	22
	Total	100

3 One temperature line on graph **E** (below) shows figures for Greece (Athens). The other is for Britain.
   a Make a large copy of the graph to show the Mediterranean figures only.
   b Write a paragraph to compare the Mediterranean summer climate with the British summer climate. Say which you would prefer and why.

E Average temperatures in Greece and Britain

- - - - - average temperature in Greece (Athens)
. . . . . . average temperature in Britain

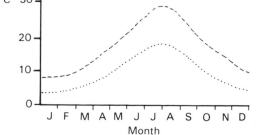

4 Study sketch **C** on the opposite page, which shows a typical hotel on the island of Corfu.
   a Write an account of how you could spend a day at this hotel. Mention the people from other countries you might meet there.
   b Find Corfu in your Atlas. Draw a sketch map to show the area within about 200 km of Corfu. Name the countries, seas, cities and other islands you can see.

Advanced work

5 Map **F** shows where the hotels on Corfu are.
   a Describe where the hotels have mostly been built. Try to give reasons why they have been built there. Use these key words in your description: coast, inland, beach, tourist, roads, main town, location.
   b Imagine you wanted to build a new hotel on Corfu. Draw a sketch map to show where you would built it. Write a report to explain your decision.

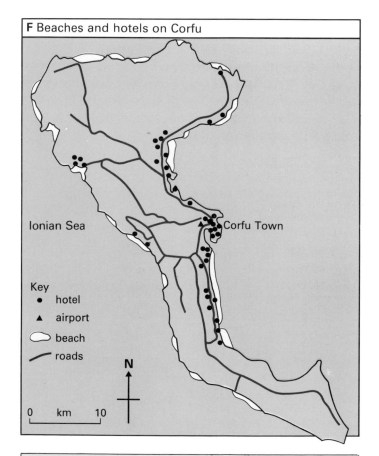

F Beaches and hotels on Corfu

Ionian Sea

Corfu Town

Key
- ● hotel
- ▲ airport
- ◯ beach
- — roads

N

0   km   10

Summary
Greece is a popular Mediterranean country for summer holidays. It is easy to get there by air. It has an environment that most holidaymakers from Northern Europe find different and attractive.

A British holiday

A

The family in sketch **A** are going on holiday. They are heading for the counties of Devon and Cornwall in south-west England.

They chose the south-west for several reasons. In summer, the weather is usually warmer than in places further north. Palm trees grow along the sea front in some seaside resorts. The family would be wise to pack waterproofs as well as sunglasses, though, just in case.

South-west England has been much easier to get to since motorways have been built. Map **B** shows how long it takes to reach the area from some of Britain's biggest cities.

There is plenty to see and do while touring in the south-west. Inland there are two National Parks, Dartmoor and Exmoor. These are highland areas with special beauty spots for picnics. Dartmoor's wild ponies are another attraction.

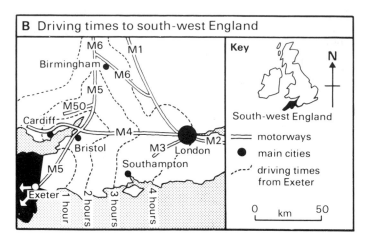

B Driving times to south-west England

Key
South-west England
N
—— motorways
● main cities
- - - driving times from Exeter
0 km 50

On the coast there are bays with sandy beaches. Small fishing villages are full of interest. Larger seaside resorts provide places to stay, amusements and summer shows.

Torquay is the number one seaside resort. It is at the back of a long bay called Torbay. The bay is sheltered from strong south-west winds, and the beach faces the sun for most of the day. It is safe to bathe in the shallow water.

All the usual seaside shops and amusements are there. Pleasure boats sit at anchor in the old harbour, though there are still some fishing boats to be seen. Pleasure cruises take holidaymakers along the coast for a breath of sea air.

C Torquay

Key
1 Hopes Nose
2 Thatcher Rock
3 Meadfoot Beach
4 Ellacombe
5 bed and breakfast guest houses
6 Wellswood
7 holiday flats
8 coach station
9 hotels
10 amusement centre
11 yacht basin
12 Haldon Pier
13 inner harbour
14 Pavillion
15 main shopping street
16 gift shops
17 Princes Gardens
18 park
19 Princes Theatre
20 outer harbour
21 Princes Pier
22 Abbey Sands
23 Torbay

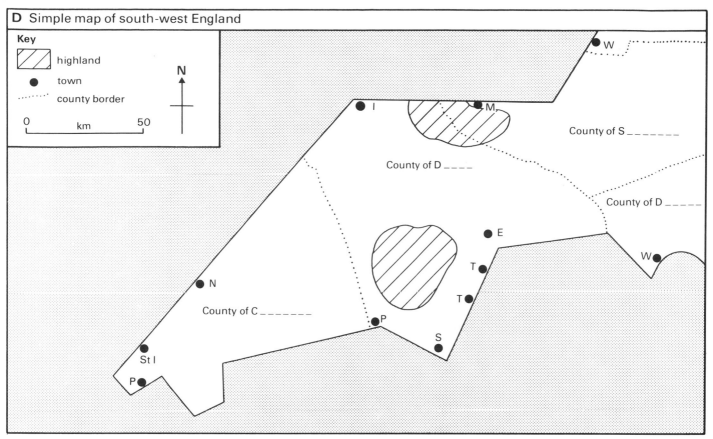

D Simple map of south-west England

Key

////// highland

● town

.... county border

0   km   50

N

County of S _ _ _ _ _ _ _

County of D _ _ _ _

County of D _ _ _ _ _

County of C _ _ _ _ _ _ _

W
I
M
E
T
T
W
N
P
S
St I
P

1 a Make a copy of map **D**, a simple map of the
south-west area.
b Use an Atlas to name the towns, counties
and highland areas.

2 Take each of the following subjects and write
a sentence about it which describes the
attractions of south-west England.
a weather
b coast
c towns and villages
d inland
e access

3 Study map **B** on the page opposite. Write out
and complete the following paragraph:
'Fast roads called .... link the south-west
to major towns and cities. Birmingham is
linked to Exeter by the M .... People from
London can drive along the M .... to ....
then head south on the M5. London and
Birmingham are within .... hours' driving
time of Exeter.

4 Write a description of Torquay seaside resort
using the information and photograph **C** on
the page opposite. Use the following headings:
Landscape and scenery
Accommodation
Entertainment

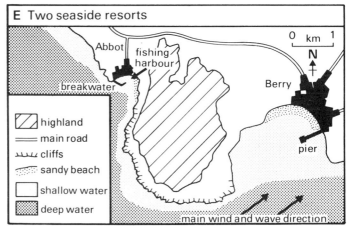

E Two seaside resorts

Abbot
fishing
harbour
breakwater
0 km 1
N
Berry
pier

////// highland
≡ main road
₩₩₩ cliffs
⸝⸝⸝ sandy beach
☐ shallow water
▓ deep water

main wind and wave direction

Advanced work
5 Study map **E**, which shows the towns of
Abbot and Berry.
a State which of the two towns you would
prefer to stay in for a holiday.
b Give reasons for your choice. Try to say
exactly what you could do there.

Summary
The south-west counties of Britain have a good
mixture of attractions for tourists. Many of
these things are the work of nature, such as the
beaches. Others are man-made, such as the
amusements. Motorways have made it easy to
get to the south-west.

Unit 8.1
Natural disasters/Floods

A

A **disaster** is an event which brings great trouble and hardship. Natural disasters usually occur because of unusual movements of earth, wind or water. One type of natural disaster is a **flood** (photograph **A**). A flood is a flow of river, lake or sea water over land which is usually quite dry. The result of a flood is shown in photograph **E**.

The worst floods of all are not necessarily the biggest. They are the floods which take people by surprise. Some predictable floods can be very helpful. (Something is 'predictable' if people can tell when it is going to happen.) Here are some descriptions of four different types of flooding.

1 Predictable river floods
The river Nile used to flood part of Egypt each July. The rest of Egypt is very dry in summer, but the Nile carries storm water from the Ethiopian mountains. Everyone waited eagerly for the floods to bring water to the parched fields. People who could predict the floods (by keeping records) were thought to have magic powers. Since 1967 the flood water has been controlled by the Aswan dam. Diagram **B** shows what happens.

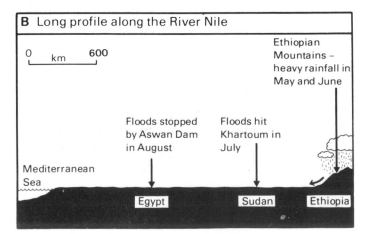

B Long profile along the River Nile

2 Unpredictable river floods
These happen when rain storms hang over a place for much longer than anyone expects. The rainwater runs off the land into streams which feed the river. The heavier and longer the storm, the worse the flooding. In the worst flood ever recorded, in August 1931, 3.7 million Chinese people were drowned by the Hwang-Ho (Yellow River). This is shown on map **C**.

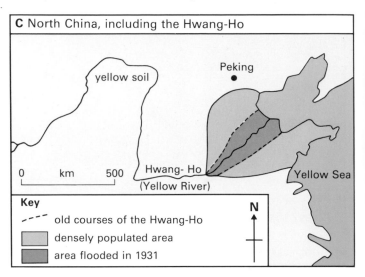

C North China, including the Hwang-Ho

3 Sea floods
In 1953, 1500 Dutch people were drowned when the North Sea flooded over the Rhine Delta. Such floods happen when the highest tides have gale force winds behind them so that the sea level is higher than anyone expects. The Dutch are making sure that floods like this do not happen again (see Unit 10.3). London is also in danger of flooding. A £500 million flood defence scheme on the river Thames was built in the mid 1980s.

Tropical storms and **tsunamis** cause sea flooding on lowland coasts (see Unit 8.2).

4 Dam bursts
The causes of dam-burst floods are shown on sketch **D**.

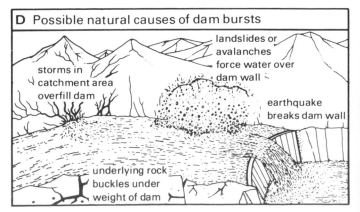

D Possible natural causes of dam bursts

1 Why are the worst floods the unpredictable ones? Write a paragraph to explain.

2 Copy the Downword **F** and fill it in from these clues.
   i The movement of water (4).
   ii This or an avalanche may cause waves which force water over a dam wall (9).
   iii The River Nile floods pass this city in July (8).
   iv A capital city which is in danger from sea floods (6).
   v Controls flood water in Egypt (5, 3).
   Read across from (i) to see if you have got it right.

F Downword

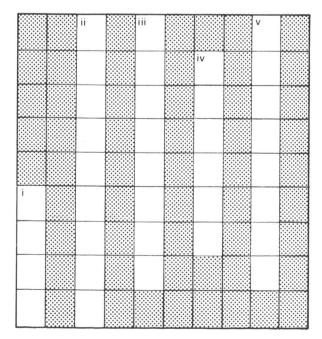

3 Draw map **C** showing the Hwang-Ho and write a fifty-word report with the title 'The Yellow River'.
Below there is a list **G** of key words which may help you.

> **G   Key words for question 3**
> Hwang-Ho, North China; colour of soil, of river and of sea; worst flood ever; how many people; people living on flood plain; sudden flow; people caught unawares; disaster; changes in river channel.

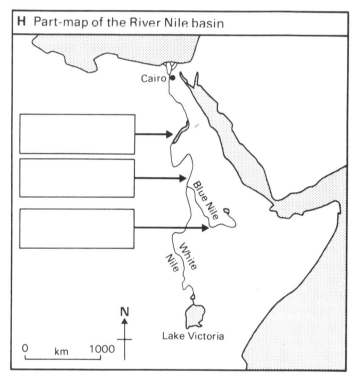

H Part-map of the River Nile basin

4 Write a short paragraph to explain how dams may burst. Use sketch **D** to help you.

5 Use the long profile **B** and your Atlas to draw a map to show how floods travel down the River Nile. The part-map **H** shows you how to set about it.

Summary
A flood is a flow of water over land which is usually dry. It may be predictable or unpredictable. Unpredictable floods can bring great trouble and hardship to people.

Pakistan fears up to 200,000 flood deaths

(Newspaper headline for a cyclone disaster)

Not much was left after the flood waters had gone down. A 25-metre-high wall of water swept in from the sea and carried all before it. Many of the people were swept away and drowned. Some were trapped in buildings then crushed as the walls collapsed. Sheet iron roofs were ripped off the houses by the wind and cut into fleeing people like circular saws.

The giant wave was caused by a **tropical cyclone**. A cyclone is a swirling mass of air about 200 km across. Wind speed is up to 200 km per hour. Rain falls in violent thundery downpours. A cyclone moves quickly over hundreds of kilometres in a day. It has a calm 'eye' in the middle.

Cyclones are formed over warm sea areas. The air becomes warm and begins to rise. Minute water droplets form on the rising air. They make clouds which hold thousands of tonnes of water. They can reach 15 km high. After a week or so, the cyclone reaches cooler seas or land. It then loses energy and dies out.

The cyclone in the headline was spotted by a US weather satellite eight hours before it struck Pakistan (now Bangladesh). At that time, it was 320 km to the south, but moving fast. There was nothing anyone could do to warn or protect the people.

There were extra high tides in the Bay of Bengal on that day. The strong winds trapped the high tide water and whipped up giant storm waves. It was these waves which caused so many deaths on the low coastal plains.

You may think that it is silly for people to live in places where cyclones happen. In this part of the Indian Ocean, there are about six cyclones every year. Since 1970, the same area has been struck by many more cyclones, some of them causing just as much damage and death.

People live here because the soils are very good for growing crops. Besides, other parts of the country are already crowded. The people must choose between not being able to grow food, which means certain death, and risking a cyclone. This is a hard choice to have to make.

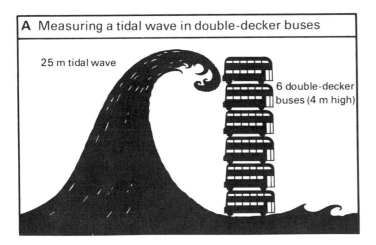

A Measuring a tidal wave in double-decker buses

25 m tidal wave

6 double-decker buses (4 m high)

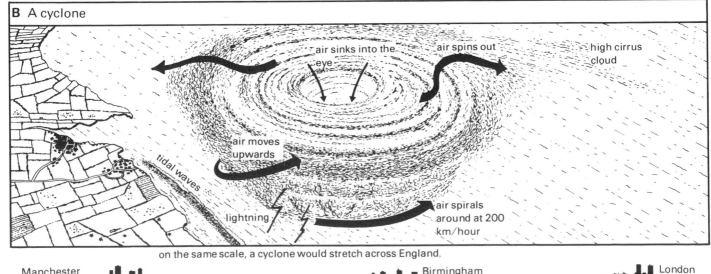

B A cyclone

air sinks into the eye

air spins out

high cirrus cloud

air moves upwards

tidal waves

lightning

air spirals around at 200 km/hour

on the same scale, a cyclone would stretch across England.

Manchester   Birmingham   London

C Cross-section through a cyclone

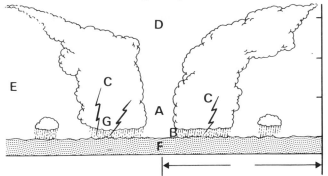

1 Make a copy of diagram **C**, which shows a cross-section through a cyclone. Write these labels by the right letters on the diagram:

A	the eye	**B**	120 km winds
C	rain clouds	**D**	15 km high
E	200 km across	**F**	warm sea
G	lightning		

D The cyclone zone

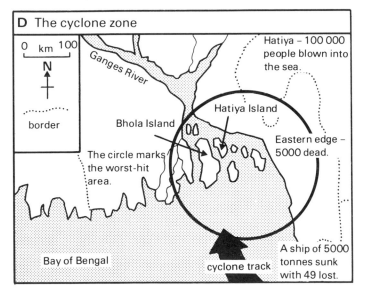

2 Study the information and map of the Bangladesh cyclone disaster. Answer these questions with a sentence for each:
   a Name the bay the cyclone came in from.
   b Which large river flows into the Bay of Bengal?
   c What does map **D** tell you about the land in the most affected area?
   d What was it that caused so many deaths during the cyclone?
   e Why did the people not have any chance of escaping the storm waves?

3 Map **E** shows other parts of the world where there are cyclones.
   a Draw a map to show where cyclones form and in which direction they move.
   b In different parts of the world, cyclones are given different names. Print these names on your map.

E Tropical cyclones around the world

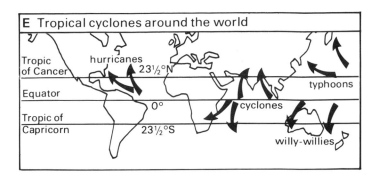

c Write a paragraph to say where most of the world's cyclones seem to start. Mention the following things in your answer:
   the latitude; over land or sea;
   hot or cold areas; the direction (track) they follow.
d Read the description of how cyclones form. Give reasons why Britain does not have cyclones like these.

F Weather satellite spotting a cyclone

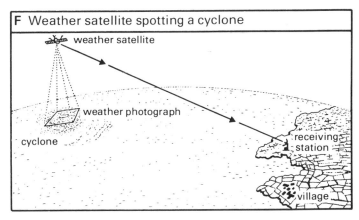

4 Study diagram **F**.
   a Say how you would run an early-warning scheme to tell people about cyclones.
   b Make a list of helpful advice you would give to people in a farming area like the one in Map **B** to help them survive a cyclone.

Advanced work

5 Imagine you lived in the area affected by the cyclone described on the opposite page. You were one of the people who survived. Would you want to rebuild your house in the same place, or would you want to move elsewhere? Describe what you would do.

Summary
Cyclones are areas of very violent weather. They are found in tropical areas. They can cause great damage when they strike populated areas. Many people risk being killed by cyclones in order to have land to grow food.

Drought

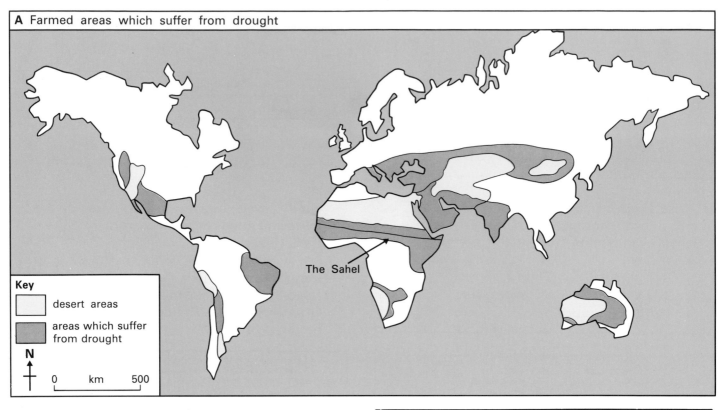

A Farmed areas which suffer from drought

Key
- desert areas
- areas which suffer from drought

N

0   km   500

Drought is lack of rain. Deserts are places which have the worst droughts. Hardly anyone lives in deserts, but many people live on the dry grasslands which surround deserts. They earn their living from growing crops or rearing cattle, sheep or goats on the dry, hot grassland. It is these folk who suffer in the great droughts.

Map **A** shows where these people live. In these grassland places it is difficult to predict when the rains will fail. When this happens, no rain falls for a long time and there is a drought.

When the rains fail crops and grass do not grow. Rivers dwindle away to nothing, animals grow thin and die. Then people starve. The worst droughts are those which go on and on for several years.

Any food stored in case of drought is eaten in the first year or so. After that people eat the seed they were going to plant – if they have any. After that they have to rely on national and international aid to survive.

The worst **famine** area in the world is the **Sahel** in Africa. This is a belt of grassland lying south of the Sahara Desert. It usually has summer rains which pour down from thunderstorms and heavy afternoon showers. In some years the wet ocean winds do not reach into the inland parts of the Sahel, and the rains fail.

B	Drought years in the eastern Sahel		(D = drought)
1960 D	1967 D	1974 D	1981 D
1961	1968	1975 D	1982 D
1962	1969	1976 D	1983 D
1963	1970 D	1977 D	1984 D
1964 D	1971 D	1978 D	1985
1965	1972 D	1979 D	
1966	1973 D	1980 D	

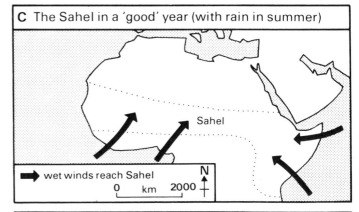

C The Sahel in a 'good' year (with rain in summer)

Sahel

→ wet winds reach Sahel

N

0   km   2000

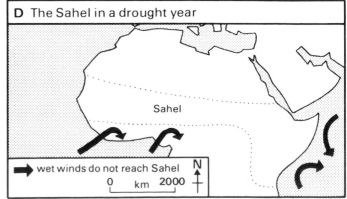

D The Sahel in a drought year

Sahel

→ wet winds do not reach Sahel

N

0   km   2000

1 Copy Map **A** opposite.

2 Look at Appeal Poster **E**. Find the countries mentioned in an Atlas. What are your reactions to the photograph?

3 Look at the list **B** of drought years in the Sahel. Write a paragraph to say which year you think would have the worst famine.

4 Use the figures in table **F** to copy and complete the rainfall graph for Timbuktu (graph **G**).

Table F	Average rainfall (mm)	Actual rainfall in 1937 (mm)
January	1	10
February	1	40
March	1	10
April	3	10
May	10	30
June	20	30
July	70	40
August	80	20
September	40	10
October	2	10
November	1	10
December	1	10
Total	230	230

Even though the total rainfall in 1937 was the same as the average, this was the year of a severe drought. Why do you think this was?

G The rainfall of Timbuktu

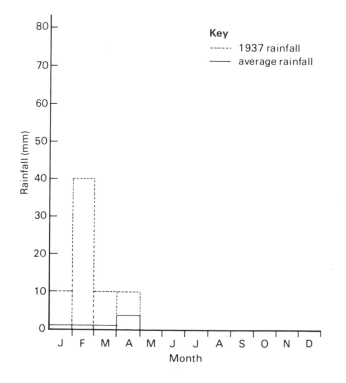

Key
----- 1937 rainfall
—— average rainfall

5 Rearrange these words so they make sense. Then try to explain what the sense is.
'It forethought much that of as famine lack as is causes drought.'

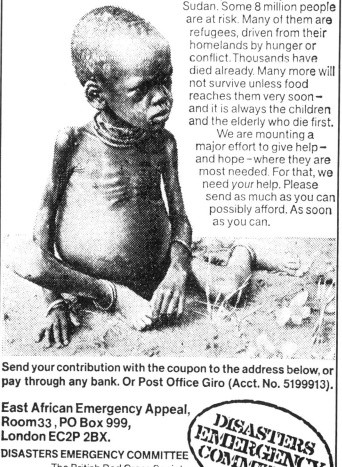

Will he ever eat another meal?

He is a bewildered victim of the worst famine in the world, which is affecting vast areas of East Africa – Uganda, Ethiopia, Somalia, Djibouti, Kenya and Sudan. Some 8 million people are at risk. Many of them are refugees, driven from their homelands by hunger or conflict. Thousands have died already. Many more will not survive unless food reaches them very soon – and it is always the children and the elderly who die first. We are mounting a major effort to give help – and hope – where they are most needed. For that, we need *your* help. Please send as much as you can possibly afford. As soon as you can.

Send your contribution with the coupon to the address below, or pay through any bank. Or Post Office Giro (Acct. No. 5199913).

East African Emergency Appeal, Room 33, PO Box 999, London EC2P 2BX.
DISASTERS EMERGENCY COMMITTEE
The British Red Cross Society
• CAFOD · Christian Aid · Oxfam
• The Save the Children Fund

DISASTERS EMERGENCY COMMITTEE

The child in this poster has a fat stomach. This does not mean he is well fed. It is a sign that he does not get enough of the right food. This is called malnutrition.

Summary
Drought is lack of rain. Deserts have the worst droughts. Drought causes most hardship to people living on dry grasslands on the edge of the deserts. The worst hit of these is the Sahel in Africa.

Unit 8.4
Earthquakes

Hope running out for El Asnam earthquake victims

From Walter Schwarz
in El Asnam

Hopes of saving more victims of the El Asnam earthquake were fading last night. Only 14 people were brought out alive yesterday when Algerian Government officials said they expected the final toll to go beyond 20,000 dead.

In the town centre, and two inner suburbs, the worst-hit areas, whole blocks are heaps of rubble. A model workers' block of flats built for 3,000 after the last earthquake here in 1954, was built with all elegance on four strories above a row of shops. Now it has caved in, burying a large number of its inhabitants.

At the flats, rescuers' efforts quicken whenever a succession of warm smells and objects signifies life, schoolbooks, bedding, toys. A high proportion of children died on Friday because many of their parents were at prayer.

El Asnam had tripled its population since the last quake in 1954, with the result that a lesser upheaval caused much

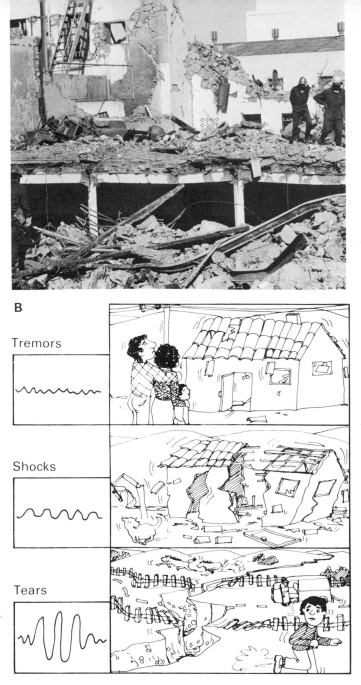

Sketch **B** shows three types of earthquakes. They are:

1 Tremors. These are gentle, slow movements of the land. They rattle cups and saucers on tables. In stronger tremors people feel the earth moving underfoot. Buildings shake, rattle and groan. Bits fall off.

2 Shocks. Sudden waves of energy pass through the land. They knock down buildings and break gas, water and sewage pipes. Fires start as sparks ignite escaping gas.

3 Tears. Parts of the earth sink, or slide sideways. Other parts shake but stay in the same place. Between the various parts the land tears. Roads and railways may be completely cut.

Earthquakes are the vibrations caused when rocks deep in the earth's crust move. These rocks are from 5 km to 1000 km below the surface, and as they move they drag the continents and ocean bed with them. Pressure builds up and up until, suddenly, the rocks jerk forward, causing an earthquake. Some earthquakes in ocean beds cause huge waves called **tsunamis** which may sink ships and cause damage to coastal areas.

Sometimes one huge area, like India, moves towards another, like Asia. All the rocks between are squashed up into **fold mountains**, like the Himalayas. This takes millions of years and causes millions of earthquakes. The crust often cracks open in such places, letting molten lava rise up to the surface. This is how **volcanoes** happen.

All this explains why earthquakes, fold mountains and volcanoes are often found together (see map **C**).

Some areas where earthquakes are common have a lot of people in them. Japan and California are two places like this. People are prepared to risk the danger of earthquakes because they like living there. Scientists think that another earthquake will hit San Francisco soon. It was nearly destroyed by one in 1906.

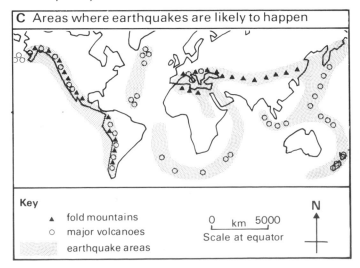

C Areas where earthquakes are likely to happen

Key
▲ fold mountains
○ major volcanoes
░ earthquake areas

0 km 5000
Scale at equator

N

Things to do

1 Copy Map **C**. It shows the areas of the world where earthquakes are likely to happen.

2 Copy out conversation **D**, adding these five words in the correct places:
shock, torn, fires, earthquake, tremor.

D 'Have you read this? It says that in Jamaica the ground heaved and sank. _ _ _ _ _ started. It stopped the cricket for half an hour!'
'I was in an _ _ _ _ _ once. Of course it was only a _ _ _ _ _. It just rattled the windows a bit. It frightened me, though.'
'You were lucky! Where I come from they still talk about the _ _ _ _ _ that knocked a factory chimney down.'
'A chimney – that's nothing! They had to bolt the factories together in my town in case they were _ _ _ _ apart. It would take more than that to stop Rugby League.'

3 Study maps **E**, **F**, **G**, **H** and **I**. Copy map **H** and fill in the missing name. Try to draw map **J**.

Don't worry if you are not sure. Make a sensible guess.

4 You are a reporter in an earthquake – scoop! Write down the news story you send in to the newspaper. Use your Atlas to put real places in.

5 Look at a map of world population distribution in your Atlas. Make a list of places where you find a high density of people in an earthquake area. What can you predict for these areas?

Summary
Earthquakes are vibrations caused by movements of the earth's crust. Three types of 'quakes' are tremors, shocks and tears. Earthquakes are linked to volcanoes, fold mountains and giant sea waves. Many people live in places which suffer from earthquakes.

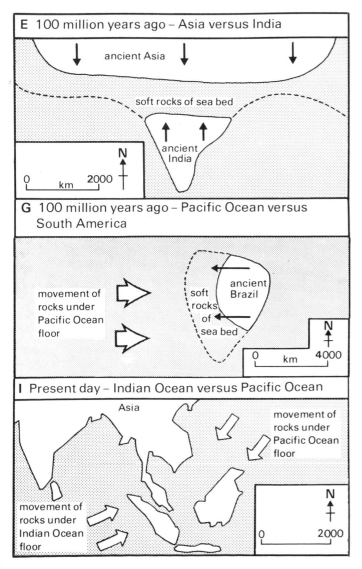

E 100 million years ago – Asia versus India

G 100 million years ago – Pacific Ocean versus South America

I Present day – Indian Ocean versus Pacific Ocean

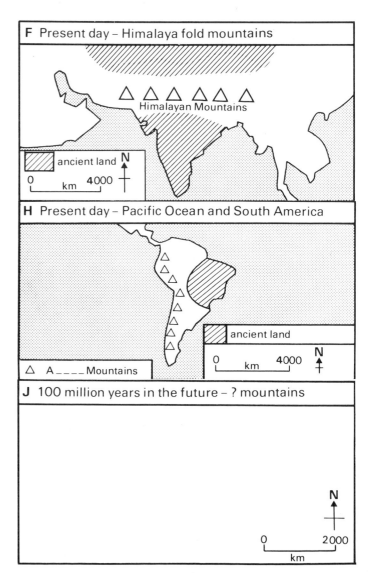

F Present day – Himalaya fold mountains

H Present day – Pacific Ocean and South America

J 100 million years in the future – ? mountains

Unit 9.1
Making a desert

A

Every year, the amount of new desert land in the world adds up to an area the size of Ireland (70 000 km²). Grasslands and forests are being turned into barren desert. (See photo **A**.)

As the number of people in the world increases, more food has to be grown. One way to do this is to plant on more land and to graze more cattle. This can destroy vegetation and soil until the land becomes useless.

In West Africa, nomadic herders graze their cattle on the grasslands. When the grass is gone, they move on. In time, the grass grows again and they can return. (See diagram **B**.)

Farmers plant crops on cleared forest land. When the soil loses its goodness because of continual planting or heavy rain, a new plot is used instead. After a few years' rest, the old plot can be used again.

These methods do not work where there are too many people to be fed. Too many cattle destroy the grass and trample the soil. This is called **overgrazing**.

Near the villages, all the land has to be used year after year without being rested. More water is needed, so wells and streams dry up.

With no vegetation and no water, the soil soon becomes dry and useless. Then the people have to move to another area which they may also ruin. Diagram **C** shows what happens.

A farmer with a family to feed does not have much choice. Food is needed every day. Problems in the future must be forgotten. Some farmers do not even understand why the land is being ruined. Fewer children would help, but there is no easy answer to the problem.

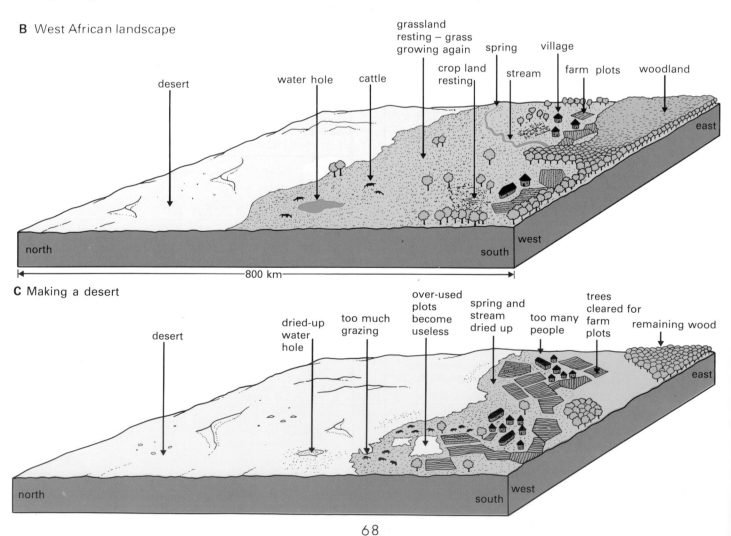

B West African landscape

C Making a desert

1 **a** Make a sketch of the photograph **A**.
  b Describe what the photograph shows.

2 Join up the 'tails' to the correct 'heads' in the following sentences:

a There is a need for more food	cutting down trees and clearing the ground.
b Farmers can get more land by	eat up all the grass.
c Too many cattle	because there are more people than there were before.
d Land farmed without being rested	when too much water is taken from wells.
e Streams and springs dry up	becomes useless in a few years.

3 Map **D** shows land which is at risk of becoming desert.

Two of the following sentences are true. Write them out.

– Most areas at risk are near present desert areas.

– The main risk areas are all in the tropics.

– The earth's desert areas would double in size if all risk areas turned to desert.

– A large part of Europe is in danger of becoming a desert.

Explain why the other two statements are not true.

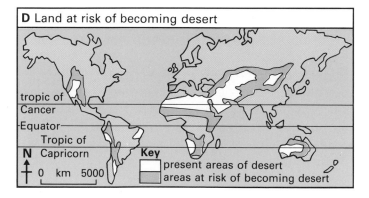

D Land at risk of becoming desert

Key
present areas of desert
areas at risk of becoming desert

4 The figures in Table **E** show how much of each continent is at risk of becoming desert. Europe and Antarctica have not been included as the problem is not serious in either continent.

Table **E**	Some risk (%)	Great risk (%)	Total % at risk
Africa	39	22	61 (example)
Asia	28	10	
Australia	63	8	
N America	13	7	
S America	16	1	

  a Draw five bars, each 100 mm high and 10 mm wide. Label each bar with the name of a continent on the chart. Each bar represents 100% of the land in each continent (all the land).
  b On each bar, mark off the **total** percentage at risk. You will have to work this out by filling in the last column on the chart. Let 1 mm represent 1%.

  c Now subdivide each bar by marking off the percentage at 'some risk'.
  d Shade in the different parts of the bars and draw a key to say what the shading means (see part-graph **F**).

F Land at risk of becoming desert

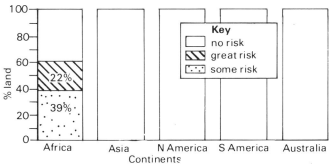

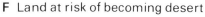

Key
no risk
great risk
some risk

Advanced work

5 Scientists helping one nomadic tribe in the African grasslands have a good way of showing how important it is to look after the land. Diagram **G** shows what they do. Describe in your own words what this 'ten tin trick' tells the tribesmen about cattle, grass and their own survival.

G The ten tin trick

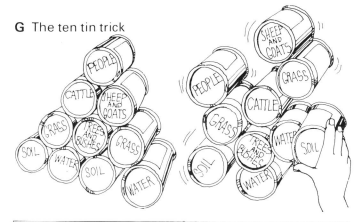

Summary
There are more people living now than ever before. More food is needed to feed everyone. People are destroying the land they depend on.

From wood to waste

A

Collecting firewood is a daily chore in poor countries all over the world. The wood is needed to cook the dinner. In some places, so many trees have been cut down that food has to be eaten uncooked.

In Asia, forests are being chopped down at a rate of 5 million **hectares** each year. This is over twice the size of Wales.

Table **B** (top right) shows why so much wood is needed and why the forests are being destroyed. It is because there are more people in Asia than ever before, and the number will go on increasing. These people have to be fed, so trees have to be cut down to provide farmland. Wood for cooking has already been mentioned.

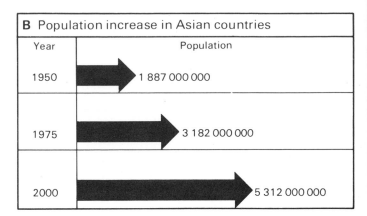

B Population increase in Asian countries

Year	Population
1950	1 887 000 000
1975	3 182 000 000
2000	5 312 000 000

Trees help protect the soil from heavy **monsoon** rains. Without protection, the soil is easily washed down the hillsides. Small valleys called **'gullies'** are cut into the bare slopes. Crops don't grow where there is no soil. The removal of soil is called **soil erosion**.

If there is no wood to cook with, something else has to be used instead. Cow dung is burnt in parts of India where there is no wood left. The dung could have been better used as a fertilizer to improve the soil.

Forest wildlife is also at risk. Some forest animals are almost extinct already. They would be certain to die out if all the forests in Asia were cut down.

Some of these problems can be solved. Gullies can be filled in. Trees can be planted to stop more soil erosion. Making an area a National Park or **game reserve** can save the animals. At the same time, the people have to eat to stay alive. Solving these problems is no easy job.

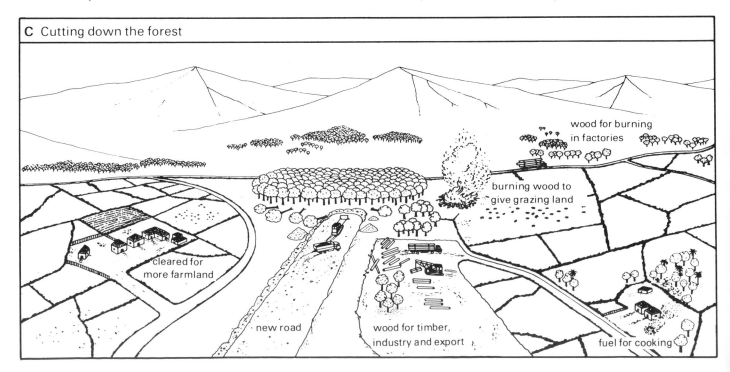

C Cutting down the forest

wood for burning in factories

burning wood to give grazing land

cleared for more farmland

new road

wood for timber, industry and export

fuel for cooking

Things to do

1 a Draw the linked boxes in diagram **D** (below).
  b Label in the reasons why trees are being felled in Asia:
  cooking – factories – grazing land – export – roads – food.

D How wood is used

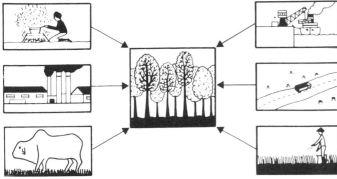

2 One thing can lead to another. Using the words below, write out and complete the following sentences to show how chopping down trees can lead to problems for the farmer.
  a Every year the .... of Asia increases.
  b Because there are more people, more .... must be grown.
  c To grow more food, more .... is needed.
  d More land can be had by felling more ....
  e Felling trees can ruin the .... for the future.

  food – population – land – trees – soil

3 Diagram **E** (below) shows what can happen when trees are felled in a tropical area. Describe these problems under the following sub-headings:
  Soil   Irrigation   Flooding   Reservoirs

E Soil erosion

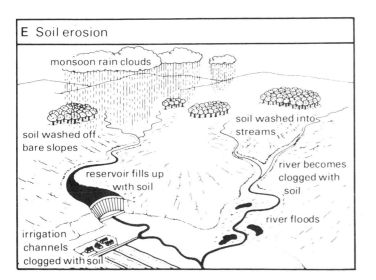

4 The animals in sketch **F** are all on the 'endangered species' list. These are animals which are very rare and which may soon be all gone (extinct). They live in tropical forest areas. Write down what you think these animals would like to say about forest clearance if only they could talk!

F Animals on the endangered species list

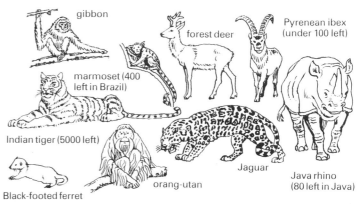

Advanced work

5 Some scientists believe that felling trees could be very serious in the future to all forms of life on earth. Diagram **G** shows why this could be true. In the last 100 years, 40% of the Amazon forest has already been felled. Some believe that the Amazon forest gives up to half the world's oxygen.
Write a short story to describe what might happen if the world began to run out of oxygen. Give your story the title. 'The day the air ran out'.

G A tropical forest

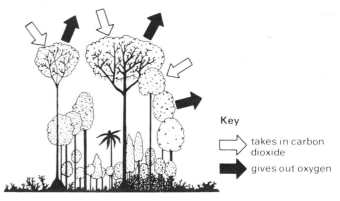

Key
⇨ takes in carbon dioxide
➡ gives out oxygen

Summary

Trees are under attack all over the world, but especially in poor countries in tropical areas. Felling trees can give extra land for more food. Too much tree felling can bring serious problems to the environment.

Unit 9.3
Polluting the seas

The fishermen of Brittany are used to seeing giant oil tankers. On 16th March 1978, the **supertanker** 'Amoco Cadiz' was heading north, bound for London. It was taking 220 000 tonnes of crude oil from Saudi Arabia to the UK. **Gale force** winds lashed the ship.

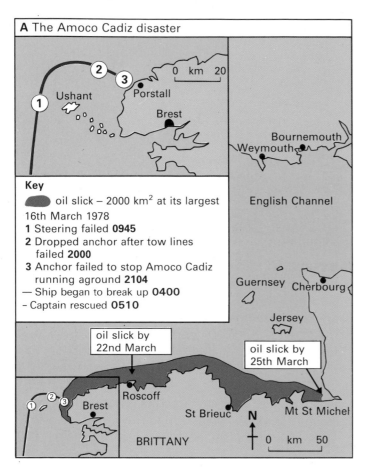

A The Amoco Cadiz disaster

Key

- oil slick – 2000 km² at its largest 16th March 1978
1 Steering failed **0945**
2 Dropped anchor after tow lines failed **2000**
3 Anchor failed to stop Amoco Cadiz running aground **2104**
— Ship began to break up **0400**
- Captain rescued **0510**

oil slick by 22nd March

oil slick by 25th March

At 9.45 a.m., the ship's steering went wrong and the supertanker began to drift out of control. A rescue ship arrived and took the Amoco Cadiz in tow, but the line broke. Twelve hours later, it ran onto submerged rocks near the coast. The gales continued, and soon the tanker split open. Crude oil from all fifteen oil tanks began to spill into the sea.

For the next two weeks, strong west winds blew the **oil slick** along the Brittany coast. Holiday beaches and fishing areas were coated with a black oily slime.

An expensive 'clean-up' job was needed to get rid of the pollution. In time, the beaches, fish and wildlife may return to normal.

There is something exciting about a disaster at sea. Newspapers and TV always report stories like that of the Amoco Cadiz. But pollution is going on all the time without being reported.

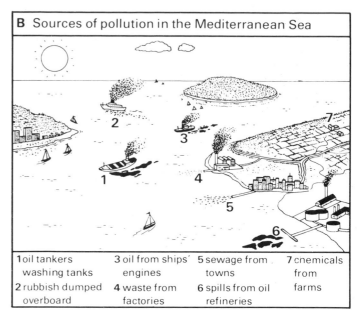

B Sources of pollution in the Mediterranean Sea

1 oil tankers washing tanks	3 oil from ships' engines	5 sewage from towns	7 chemicals from farms
2 rubbish dumped overboard	4 waste from factories	6 spills from oil refineries	

More tourists swim in the Mediterranean sea than in any other sea in the world. It is also the world's most polluted sea. Between half a million and a million tonnes of oil are spilled into it every year. It comes from tankers washing their tanks at sea, accidents at oil refineries, and simple leaks.

At the same time, raw sewage is being emptied daily into the Mediterranean from towns and cities all along the coast. Factories use the sea as a dumping ground to get rid of waste. Farmers add to the problem when rain washes chemical fertilizers or insect killers off their land into the sea.

Seas and oceans can break down pollutants and make them harmless. But too much pollution uses up all the sea's oxygen. With no oxygen, the fish die.

Some firms and governments do try to stop pollution. But stopping pollution costs money. If we want clean water we must pay the cost of keeping it clean.

Did you know that:
- The hull of a supertanker, compared to its total size, is thinner than the shell of an egg compared to the size of an egg.
- Every day, somewhere in the world, a ship is lost at sea.
- More crude oil is shipped over the oceans than any other commodity.
- There are 7000 oil tankers in the world.
- The Mediterranean takes up less than half of 1% of the world's water. But it has between 10% and 25% of the oil pollution.

Things to do

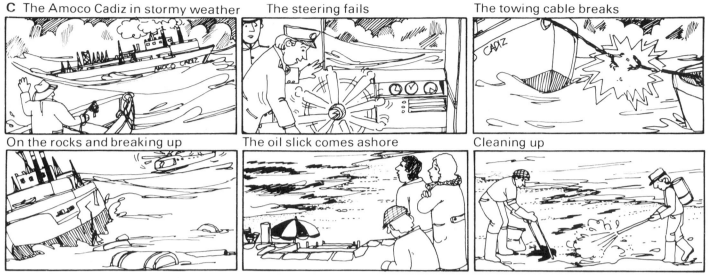

C The Amoco Cadiz in stormy weather | The steering fails | The towing cable breaks

On the rocks and breaking up | The oil slick comes ashore | Cleaning up

1 Use diagram **C** and map **A** to write an account of the events which led to the wrecking of the Amoco Cadiz. Write it like a newspaper report and invent a suitable title.

2 Study map **A** and the information on the opposite page. Answer these questions with a sentence for each:

   a Name the part of France which was affected by the oil spill from the Amoco Cadiz.

   b What is the name given to the patch of oil which floated along the coast?

   c How many kilometres did the oil stretch along the coast? (You may measure in a straight line.)

   d Why were people in seaside resorts so worried about the oil spill? (Hint: what month did it happen in?)

   e How would local fishermen be affected by the oil spill?

3 Which of the following people do you think should have paid for cleaning the beaches? Arrange the list in the order you think right, and say how much of the total each should have paid.

   The ship's captain and crew
   The company which built the ship
   The ship's owners
   The oil company which owned the oil
   The French government
   French fishermen and hotel owners in Brittany
   Everyone living in the affected area
   (Note: money from the ship's insurance is not enough to pay for cleaning up the pollution.)

4 Study diagram **B** showing how pollution gets into the Mediterranean sea.

   a Make a list of the reasons why pollution is so difficult to stop.

   b Which types of pollution could be stopped most easily? Say how this could be done.

   c Pollution which gets into the Mediterranean takes 80 years to get out again. Study an Atlas map and explain why this is true.

   d Give some reasons why it is important to stop pollution getting into the Mediterranean sea.

Advanced work

5 Map **D** shows the main oil tanker routes around the world.
   On a sheet of tracing paper or a world outline, shade in the areas of coast which you think are most at risk from oil pollution. For extra information about winds and currents, use your Atlas.

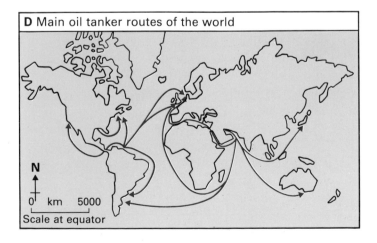

D Main oil tanker routes of the world

N

0 km 5000
Scale at equator

Summary
Pollution sometimes happens because of an accident. Sometimes the pollution can be carried a long way. Pollution is going on all the time. We must be prepared to pay to stop it.

Derelict land

There is sure to be a patch of land near your house which nobody seems to own. It looks uncared-for and people dump rubbish on it. Children play there, though it may be fenced off and dangerous.

There are thousands of places like this in Britain. Some are only a few square metres in size. Others are much larger. The word **'derelict'** is used to describe land like this. The total amount of derelict land in Britain adds up to more than the size of several large cities put together.

In towns, the derelict land is usually near the centre. Diagram **B** shows some reasons why land becomes derelict and what happens to it.

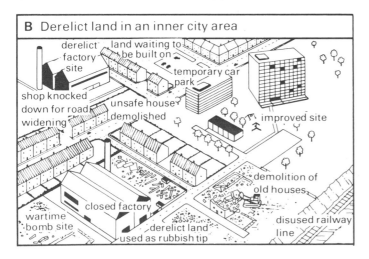

B Derelict land in an inner city area

There is also derelict land in the countryside. Mines and quarries can ruin a green landscape. Old buildings fall into ruin and become eyesores.

There is now a law which makes sure that land which has been used for mining and quarrying is restored. It must be cleaned up and put back to its original use, or put to some other good use.

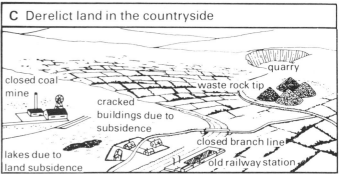

C Derelict land in the countryside

In the Tawe valley in South Wales, 200 years of mining and industry had ruined the landscape. There was 7 million tonnes of waste from factories which had made copper, zinc, tin and steel. Waste tips took up 3000 hectares of land (see map **D**). There were also ruined and crumbling buildings.

Much of the area has been cleaned up now. Waste tips were bulldozed flat and grass was planted. Local schoolchildren helped to plant trees to make a Forest Park along the river Tawe. The area looks much more pleasant now and people can enjoy picnics on the restored land.

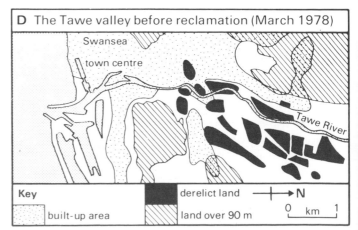

D The Tawe valley before reclamation (March 1978)

Key
built-up area
derelict land
land over 90 m
N
0   km   1

1 Think about a patch of derelict land you know.
(If you do not know any derelict sites, describe
fully the site shown on the photograph **A**.)
  a Say where the derelict land is or draw a map
  of its location.
  b Describe what is on the site now.
  c Say what you think was on the site before
  it became derelict.
  d How long do you think the site has been like
  this?
  e What is the site used for now?
  f What do you think would be the best thing to
  do with this site now?

2 List the following words and give each its
correct meaning:

derelict	an ugly sight
restore	land which was once used but is now waste
demolition	the sinking of land over mine workings
eyesore	knocking down (buildings)
subsidence	to clean up and put to use again

3 Study diagram **B** which shows derelict land
in a town. Map **E** shows the same area.
  a Make a copy of the map outline.
  b Label in what is to be found at each of the
  sites numbered 1 to 11 on the map.
  c Write a paragraph to say whether you would
  like or dislike to live in an area like this, and
  explain why.

F A coal mine in a South Wales valley

G The same valley after restoration

4 Photographs **F** and **G** show a landscape before
and after it was restored.
  a Say how the land was used before it was
  restored.
  b Draw a sketch of the second photograph (**G**).
  Label all the things which have changed.

Advanced work
5 a Why do you think land was not restored in
  the past by mining companies and factory
  owners?
  b Give some reasons why land should be
  restored after use.

Summary
Derelict land is land which is dirty and unused.
It occurs in both town and country areas.
There is no need for land to be left derelict.
Present laws try to make sure this does not
happen any more.

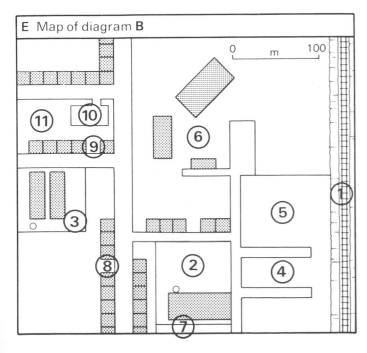

E Map of diagram B

0   m   100

⑪ ⑩ ⑨ ⑥ ① ⑤ ③ ⑧ ② ④ ⑦

Unit 10.1
The Amazon forest

Some people think that governments should be in charge of new industry. Other people think that businessmen should be allowed to build up industries in their own way, so that they risk their own money and not the government's money.

In Brazil, the government allowed an American, Mr Daniel K. Ludwig, to develop parts of the Amazon tropical rain forest in his own way. He spent £1000 million on the scheme. The idea was to make wood pulp (for paper) from trees grown near the River Jari in Brazil. This area is shown on map **A**.

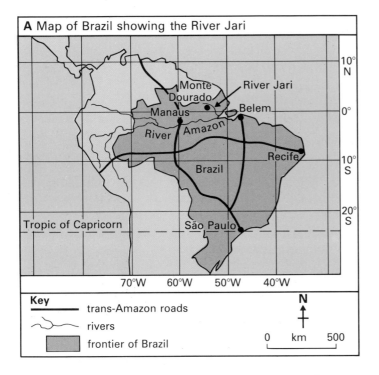

A Map of Brazil showing the River Jari

Key
— trans-Amazon roads
~ rivers
▨ frontier of Brazil

N
↑

0   km   500

This area is part of the greatest tropical forest on earth. In its natural state it is teeming with life and death. Trees and animals fight for survival and the floor of the forest is littered with decaying plant life. It is changing rapidly.

Mr Ludwig's firm have cut down some of the natural forest and have planted **gmelina** and tropical pine trees which grow quickly. Cattle are allowed to graze in the plantations to keep the forest floor clear. A **pulp mill**, 30 000 tonnes in weight, was built in a Japanese shipyard and floated (pulled by tugs) across the Pacific, Indian and South Atlantic oceans to its site on the banks of the river Jari (see photo **C**, plan **D** and diagram **E**).

One town, Monte Dourado, and five villages were built to house forest workers. Most of the tree cutters come from north-east Brazil (between Recife and Belem; where jobs are scarce. About 6000 people settled in the Monte Dourado area (see map **A**).

Further south in Brazil, things were not so organized. Along the new roads pioneers were destroying the forest and leaving ony rough pastureland or scrub in its place.

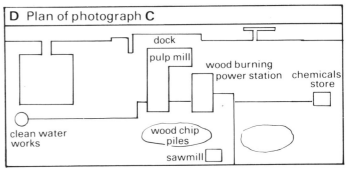

D Plan of photograph C

dock
pulp mill
wood burning power station   chemicals store
clean water works
wood chip piles
sawmill

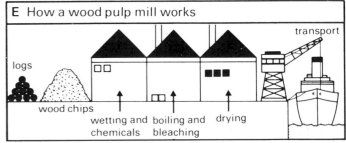

E How a wood pulp mill works

logs
wood chips
transport
wetting and chemicals   boiling and bleaching   drying

B The River Jari in 1977

C The River Jari in 1980

1 Copy map **A** of Brazil showing the River Jari.

2 Answer the following questions with one sentence each. Use diagram **F** to help you.
a What is the latitude of the mouth of the Amazon?
b What is the longitude of the mouth of the Amazon?
c What is the longitude of Manaus?
d What is the latitude of Monte Dourado?
e What are the latitude and longitude of Belem?

F How to find latitude and longitude

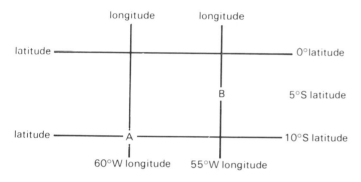

Place A is at 10°S (latitude) 60°W (longitude)
Place B is at 5°S (latitude) 55°W (longitude)

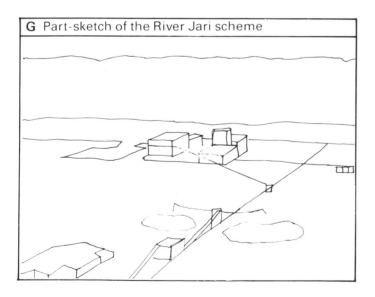

G Part-sketch of the River Jari scheme

3 Draw a sketch of photograph **C** opposite. Use part-sketch **G** to help you. Add the following labels in the correct places:
   sawmill
   woodchips
   water cleaning plant
   pulp mill
   wood-burning power station

4 a Write a half-page account of Mr Ludwig's scheme. Say why you think he chose this particular part of Brazil.

b Look at the sketches **H**. Say which of the three options you think is best for the Amazon rain forest of Brazil, and why you think so.

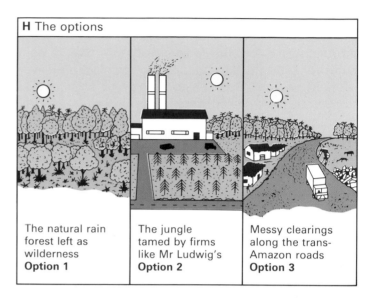

H The options

The natural rain forest left as wilderness
Option 1

The jungle tamed by firms like Mr Ludwig's
Option 2

Messy clearings along the trans-Amazon roads
Option 3

5 Mr Ludwig gambled £1000 million on this scheme. It would not have shown a profit until 1990 or so. It was a big risk for his firm. The money he spent is called **risk capital**.
List the risks. Below there are some clues (**I**) to help you.

I The risks

tree diseases

fire

floods

people diseases

hazards of the journey

Summary
The Amazon Jungle is being cut down. Along main roads this is being done badly. Along the river Jari, people are risking their money to develop the land well. Some people would like to see the jungle left as it was.

Unit 10.2
City in the desert

Saudi Arabia (map **A**) is a huge desert country which is rich in oil. It had few roads, hospitals or factories before oil was discovered. Now the Saudi government is using the money from its oil sales to build these things. For instance, a large airport has just been built at Abha (photo **G**) in South-west Arabia. At present this is an empty mountain area. Another scheme is to build a large petrochemical works, town and Red Sea port at Yan bu'al Bahr. Oil for refining and chemicals will be pumped 1200 km from the Persian Gulf oilfields.

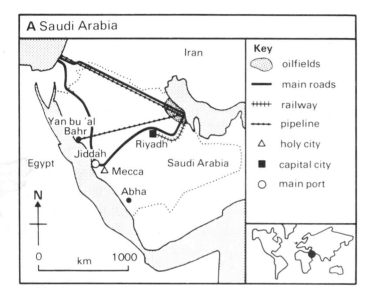

A Saudi Arabia

B The surveyors

The planners, architects and engineers

The builders

Building new towns in empty places is costly. The hope is that new jobs will attract people who will work hard to make the area rich. Until recently, most young Arabs lived a nomadic life, herding the family's sheep, goats and camels. Now the desert is changing rapidly (see sketches **B**).

First came the **surveyors,** mapping out the land. Then the planners, architects and engineers compared the maps with aerial and satellite photographs and made plans for the new city. It may look like sketch **C**. They want the city to be a base for opening up the unused parts of the country.

The builders moved in next, laying roads and building houses for the workers. Many have come here to work from other countries. Some are from Britain, France, West Germany and Italy. Most are from neighbouring countries like Egypt and Iraq.

In ten years' time this area will have the things shown on sketch **C** and listed below.
1 A new port and new roads to bring in raw materials and take out factory-made goods.
2 New factories to make things such as bicycles and TV sets.
3 New schools and technical colleges to give people skills.
4 Piped water for drinking, washing and industry.
5 Drip irrigation from piped water to grow salad crops in fields near the city.
6 Electricity brought in by cable for power.
7 New offices for business.
8 New houses for people to live in.
9 New parks for people to enjoy.
10 Entertainments such as theatres, cinemas and restaurants for leisure.

C A new desert city

1 Use map **A** and the small world map to help you to draw your own maps to show Saudi Arabia.

2 Copy out the clues below and fill in five important place names from this region (to check – find the correct answers on Chart **D**).
   a Saudi Arabia's main port is ....
   b The capital city of Saudi Arabia is ....
   c The new town on the Red Sea coast is ....
   d The oilfield is on the coast of the ....
   e An airport has recently been built at ....

D Find the words

```
P J E W X C I R G R Z I
Y L I S Y E J M O I L R
A C D D E S E R T Y A O
N D F I D K L H Q A Y P
B C G C A A R I E D R O
U B H E B M H J N H E N
A R A B C N O A L M N K
L A I E D P I A B C C J
B Z J R E S F Z D T E A
A V K G R L O Y E F B B
H X L E U R S X H T G H
R W P G T U V W G H I A
```

3 Write a story as a young man living in Arabia today might tell it to his grandchildren in the 21st century.
   Start with his life as a shepherd. Then tell of the building of the town. Finally write how he moved into a new job and home in the city. What has he enjoyed? What does he regret?

4 Design your own city. Look at sketch **E**, which shows a desert area. Part-sketch **F** shows what it may become. Use sketch **F** to add on your own new city in the desert. Label ten changes which have been made to the area.

E

F

Advanced work
5 Think out what might happen if the new towns do not work. What could go wrong?

Summary
The country of Saudi Arabia is changing rapidly. It has lots of money to spend from the sale of its oil. It is building new towns to open up empty parts of the country.

G

Land and sea

In 1421 the North Sea broke through a sandy part of the Dutch coast. The 250 000 hectares of flooded land was called the Zuider Zee (see map **A**). It lies in Holland, the northern part of the Netherlands. At 1 p.m. on 28th May 1932 this 'Zee' was dammed by a dyke. Dutch engineers then set to work to pump away the salt water and make the muddy ground fit for farming. The new farmland created this way is called **'polder'** (see sketch **B**).

The draining of the Zuider Zee is only one of the Dutch people's determined efforts to improve their coastline. Their biggest present-day project is the River Rhine Delta plan, which is shown on Map **A**.

The idea here is not to gain new land, but to defend Holland against further North Sea flooding. The scheme will also improve the rivers for shipping and will provide roads to link up the islands. It will give supplies of fresh (river) water to the coastal cities and towns.

Further north the River Lek, a branch of the great River Rhine, reaches the sea. Here Europe's greatest port has been built on land reclaimed from the sea. It is called Europoort, and it is shown on Map **A** and photograph **C**. Goods from inland industrial areas like the Ruhr valley in West Germany come to Europoort by barge and are loaded onto ocean-going vessels.

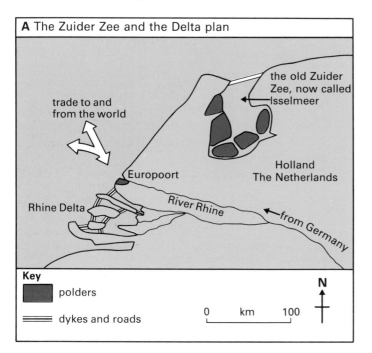

A The Zuider Zee and the Delta plan

the old Zuider Zee, now called Isselmeer

trade to and from the world

Europoort

Rhine Delta

River Rhine

from Germany

Holland
The Netherlands

Key

polders

dykes and roads

0   km   100

N

C

The future may be even more dramatic. Sketch **D** shows a cross-section of a city of glass and concrete which engineers say is possible. It would be built in the North Sea, standing like an oil rig on the sea floor. This plan was designed by an English firm. Dutch engineers have their own designs.

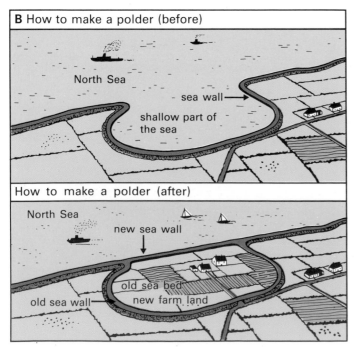

B How to make a polder (before)

North Sea

sea wall

shallow part of the sea

How to make a polder (after)

North Sea

new sea wall

old sea bed

old sea wall

new farm land

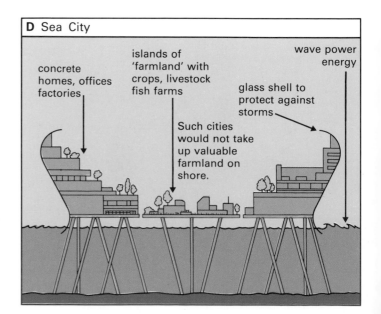

D Sea City

concrete homes, offices factories

islands of 'farmland' with crops, livestock fish farms

wave power energy

glass shell to protect against storms

Such cities would not take up valuable farmland on shore.

1 Copy map **A**, which shows the changing coast-line of the Netherlands.

2 Rearrange the phrases shown below so that they say something sensible about the Netherlands and the North Sea. When you have them correct in rough, copy them up.

    bringing death and destruction to Holland.
    The sea has also been an enemy,
    The sea has been Holland's friend,
    have made this country prosperous.
    because Dutch ships, taking Dutch, German
      and Swiss goods to the world,

3 **a** Describe how a polder is made.

  b The River Rhine Delta plan, shown on map **A**, is not meant to gain new farm land. Write fifty words to say what it is for. Use an Atlas to help you find where the dykes, bridges and roads are.

  NB: a 'dyke' is a sea wall in Holland, not a ditch as we say in some parts of Britain.

4 **a** Use the part-sketch **E** below to help you draw your own simple sketch of the photograph of Europoort.

  b Write a paragraph to explain what seems to be happening in the photograph.

Clue: there are ocean-going ships and barges in the same docks.

5 Study sketch **D** of Sea City. Try to think what life would be like on such a city. Write an essay with the following title: 'Sea, sun and sabotage on Sea City'.
(A vivid imagination is needed!)

Summary
The Dutch people are experts at 'taming' their coastline. They have turned part of the sea bed into farmland and another part into Europe's main seaport. Now they are improving the delta of the River Rhine.

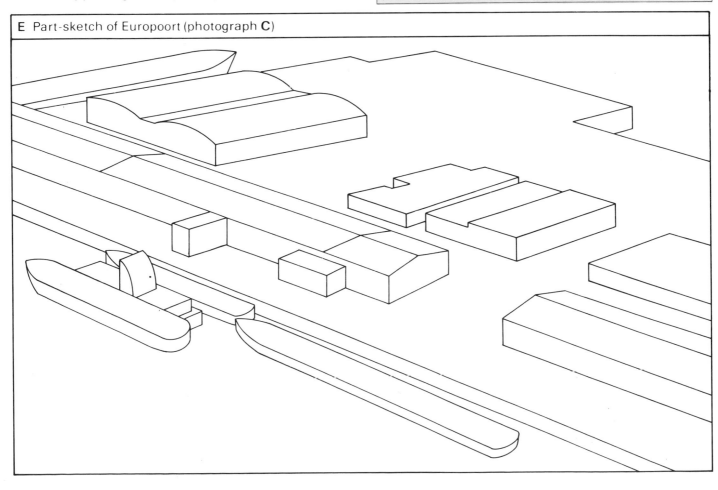

E Part-sketch of Europoort (photograph **C**)

Oregon

Oregon is a state of wind, water and wild mountains. Oregon is also a state of mind.

Oregon is one of the most beautiful parts of the United States of America. It has two mountain peaks 3000 metre high within 50 km of its capital city, Portland. It also has a gorge (Hell's Canyon) deeper than the Grand Canyon. Oregon residents are foresters, farmers, fishermen or factory workers.

Until recently Oregon suffered from dirty rivers, burned-out cars by roadsides and litter in the streets, like many other parts of the USA and Britain. Since 1970, though, things have changed. Now the people of Oregon are trying hard to care for the **environment** in which they live. (Environment means 'surroundings'.)

Chart **B** shows some of the many things which people of Oregon have changed. The new laws have improved the quality of life in Oregon. This means that Oregon is a nicer place to live in than it was before the changes.

A lot of the credit is due to a politician, the Governor of Oregon. He has persuaded other politicians, factory owners and the man in the street that it is worth a little trouble to keep Oregon beautiful. (A politician is someone who runs public affairs.)

All these ideas have been suggested in Britain. They have worked in Oregon because laws have been passed and people now pride themselves on keeping their state clean. Advertising posters like the one below (**C**) encourage people to keep on keeping on.

Nature recently added its own changes to this landscape. In May 1980 Mount St Helens, a volcano, erupted. It sent volcanic dust 15 000 metres into the air. Now tourists go to see the new shape of the mountain. Mount St Helens is in Washington State, just 50 km north of Portland.

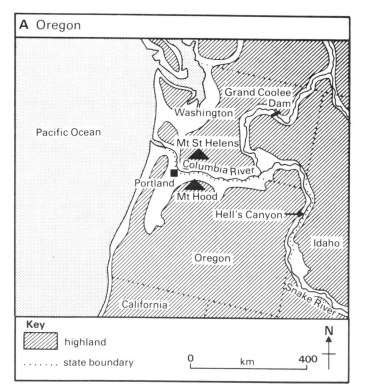

A Oregon

C

B Chart of ideas

Before	Law	After
1 Many cans were thrown away.	Cans with rip-off tops were banned.	Few cans are thrown away.
2 Broken bottles were found everywhere.	All bottles have a refund on return.	Bottle litter is rarely seen.
3 Timber mills used to pump waste into rivers.	Severe fines for dirtying rivers.	Cleaner rivers. More fish.
4 Many faulty car exhausts, giving fumes.	Car exhausts must have regular checks.	Less fumes, healthier roads.
5 Many traffic jams.	Cheap buses and trams.	No jams, cleaner cities.

Things to do

1 Copy the map of Oregon (**A**) noting the position of the main city, Portland, in the centre of the largest lowland area. You may also wish to copy map **D**.

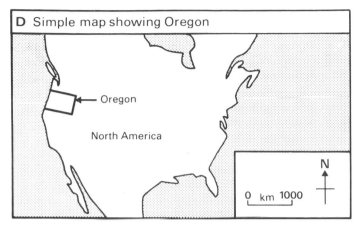

D Simple map showing Oregon

Oregon
North America
0 km 1000
N

2 Look at chart **B**. Write a sentence for each of the following, saying what happened before the law and what happens now:
   a rivers
   b traffic jams
   c cans
   d car exhausts
   e bottles

3 Many tourists visit Oregon each year. Write out a travel agency report on how to spend a holiday in Oregon. Use the plan shown in **E** and write a sentence or two on each part.

E   Plan of report on Oregon
   a Why come to Oregon?
   b A tour to look forward to!
     (i) Portland, State capital . . .
     (ii)
     (iii)
     (iv) } Do these yourself.
     (v)
     (vi)
   c Why you will want to come again.

4 Look at the diagrams **F**. Write a report on 'The rapid changes in Mount St. Helens'.

5 Can you think of anyone who might be against Oregon's new laws? Explain why.
   Clues:
   Returnable bottles mean more work for manufacturers.
   Pumping waste into rivers is cheaper than getting rid of it in a clean way.
   Motorists have to pay to have their car exhaust repaired.

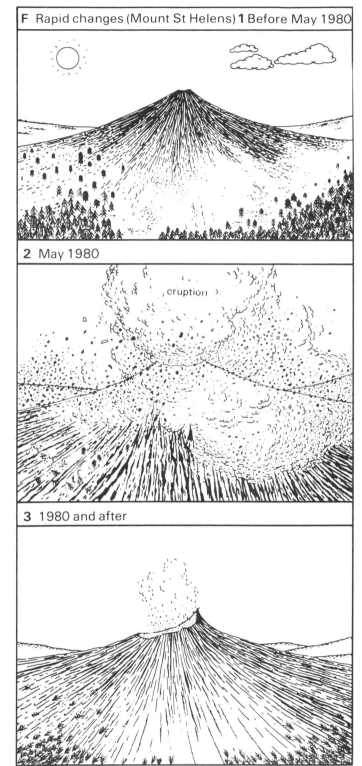

F Rapid changes (Mount St Helens) **1** Before May 1980

2 May 1980

eruption

3 1980 and after

6 Invent a poster to help improve the quality of life in Oregon.

Summary
The people of Oregon have cleaned up their state. This has improved the quality of everybody's life. They have set an example for the rest of the world.

Unit 1
Recap

Geography is the study of places and the people who live in them. Some places have few people because they are too cold, or too dry, or too steep. People can live there, but only at a high cost.

Choose the word to fit the descriptions below from this list:

> place . . . mountain chains . . . landscape . . . deserts . . . tundra

1 Places where there is little rain.
2 The polar lowlands where ice melts in summer.
3 Lines of mountains linked to each other.
4 Scenery.
5 An area with its own special character, made up of the land and people.

Give the correct term for each of these things from the list:

> ice cap . . . mountain . . . contour lines . . . freezing point . . . oasis

6 0° Centigrade.
7 Lines on a map which show how high the land is.
8 Polar layer of ice covering the ground all through the year.
9 A water-hole which allows people to live in the desert.
10 A highland over 300 metres above nearby lowland.

Name the places which fit these descriptions:
11 The great ice-bound southern continent.
12 The largest desert in Africa (and in the world).
13 The mountain chain linking Alaska with Mexico in North America.
14 The Pole surrounded by the Arctic Ocean.
15 The country in which Milford Sound is found.

Fill in each of the blanks with the most suitable word for the sentence.
16 It is too c_____ for people to live comfortably in Polar regions.
17 Most highland people live in sh_____ places
18 If farmers i_____ the desert, crops can be grown.
19 Climbers cross the s_____ l_____ when they move from rocky slopes on to permanent snow cover.
20 The deserts of the Middle East are famous for the o_____ in their rocks.

The survival game
Engine failure . . . parachutes . . . group scattered by wind . . . lost in strange countryside . . .

Can **you** survive?

Look at the board on the inside front cover of this book. You have landed at one of the points marked 'landing'. You have to get to the safe point **S**.

Decide which 'landing' square you are on. Make a paper counter and put it on that square.

Use the number circle for your moves. Pick any number on the number circle for your first move. If several people are playing, they should all choose different numbers to start on.

For example: suppose you start with the 6 at the top of the circle.

Move your paper counter 6 squares towards the safe base. Then take the next number clockwise (7), then the next number (1), and so on. (You can also choose your numbers anticlockwise as long as you stick to one direction.)

Spots marked **T** are 'trouble spots'. If you land on one, take your next turn backwards.

Spots marked **L** are 'lucky breaks'. If you land on one, move double on your next turn.

To survive, you must get to the safe base in 10 moves or fewer.

The game takes about 5 minutes on your own, or 10 minutes with two of you.

Number circle

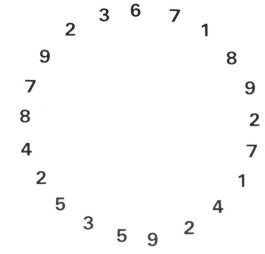

Most people in the world live on farms. A few are rich and can afford machines to do the work. Most other farming families grow their own food but have little left over to sell.

A

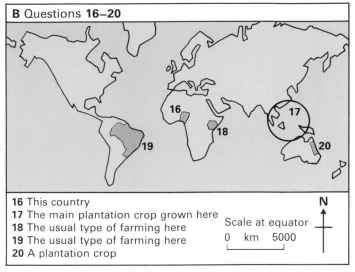

B Questions 16–20

16 This country
17 The main plantation crop grown here
18 The usual type of farming here
19 The usual type of farming here
20 A plantation crop

Scale at equator

0   km   5000

N

Sketch **A** is labelled **1-5**. Which of the labels below should go with each number?

> subsistence farm . . . manyatta . . . irrigation . . . plantation . . . nomads

Answer questions **6-10** with a word from this list:

> milk . . . yam . . . tropical . . . Nile . . . rice

6 What is the main food crop in Indonesia?
7 In which countries are most plantations found?
8 What is the main food of the Boran tribe?
9 What is the river which irrigates Egypt?
10 What is the main root crop of West Africa?

Find one word or phrase to fit each of the descriptions in brackets.
11 Farming which is (old-fashioned and learned from parents and grandparents).
12 A (quickly-made camp in the desert for young men from the tribe).
13 (Growing crops in one place, then moving on to grow next year's crop in a newly cleared forest area.)
14 A (root crop like a potato grown in tropical countries).
15 Farming (with lots of people working hard all year round in the fields).

Questions **16-20** are shown on map **B**.

The subsistence farming game
Can you feed your family through a year without having to ask for help?

How to play
Look in the inside back cover of this book. You will find a number circle. The numbers tell you how much rain will fall in each month. The more rain, the more food will grow.

Start from any number you like. (It is best to start with a high number.) After that take every **13th** number going clockwise. There are two charts on the inside back cover. Chart **A** tells you how much rain and how much food each score will bring. For example, a score of 1 gives 5 cm of rain and 200 kJ (kilojoules) of food.

It will help you to play if you copy out chart **B** and fill it in as you go along.

To start with, you belong to a family of six, including yourself.

Choose your first number and fill in the chart for January. Share out the food and decide if you are going to store any in case next month is dry.

For example, if your first number is 9, that means 45 cm of rain and 1800 kJ of food. You could give out 200 kJ each and store 600 kJ for next month.

If you come to a circled number, follow the instructions in the box. Some of them mean that you have more or fewer people to feed.

Starvation level is 120 kJ a month. If you have less than this for each person for three months, you have to ask for help.

Unit 3
Recap

Ways of growing crops and rearing animals have been improved by scientific discoveries. Modern farming is more like an industry than traditional farming. Western Europe and North America produce most of the world's food surpluses from modern factory farms.

Find the cereals which five **superseeds** came from, using the following clues:

Clue	Growing conditions	Where it grows
1 (_____ sugar?)	cool/dryish	USA, Europe
2 Nice?	hot/wet	SE Asia mainly
3 Corn	warm/wettish	USA, USSR, India
4 Dry feet?	warm/dryish	USA, USSR, India
5 Toothache?	hot/dry	Africa, India

All these things could be found on farms. Find the correct names for them from the list:

> fodder . . . drip irrigation . . . commodity . . . status symbol . . . crop rotation

6 Something you use to show off how rich you are!
7 Anything traded in its raw state.
8 Crops used to feed animals over the winter.
9 A system of changing the crop grown in each field from year to year.
10 Using plastic pipes to feed water to crops.

Give the correct farming term to fit each description.
11 The grain with the greatest output each year in the world.
12 Animals reared indoors in pens.

13 Fields left unused for a year to recover.
14 The type of farming which has animals and crops on the same farm.
15 Chemicals added to the soil to help crops to grow.

What? Which? How much? How many?
16 . . . is the name of the 'belt' of rich farming land in the USA?
17 . . . is ruined in storage, out of every 10 tonnes of food produced in the world?
18 . . . is the type of farming in which few people work huge farms with expensive machinery?
19 . . . sheep are there in the world? (10 million, 1 million, 100 million or 1000 million)
20 . . . is the grocery name for the seed-bearing part of the maize plant?

A Corn Belt farm

a Complete and label the part-sketch of photograph **B**. It shows a Corn Belt farm.
b Write an account of a day in the life of Hiram K. Corncrake, who lives on this farm.

A Part-sketch of photograph **B**

B

We use resources to make useful things which people will buy. Resources have to be found and developed before they can be used. Resources can be solid, like iron ore, or difficult to describe, like human skill.

Questions **1-5**: name the resources labelled **1-5** on sketch **A**. Choose from this list:

water . . . oil . . . granite . . . timber . . . beautiful scenery

A Types of resource

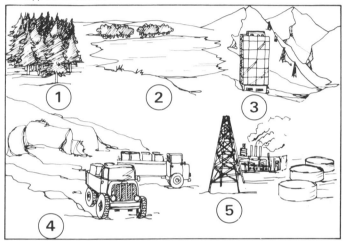

6 How many tonnes of coal are equal to one tonne of uranium?
7 Is mahogany a soft or hard wood?
8 What fuel does a nuclear power station use?
9 What is wood pulp used to make?
10 Which resource is top of the top ten for the amount produced?

Give the geographical name for each of the following:
11 Someone who wishes to live with nature, not destroy it.
12 A scientist who studies rocks.
13 Something which people can use.
14 A resource which will run out.
15 People who visit a place on holiday.

Name the important resource for which these countries are famous:

Country	Clue	Resource
16 South Africa	A metal	
17 Saudi Arabia	A fuel	
18 Sweden	A mineral	
19 Canada	A cereal	
20 Thailand	A cereal	

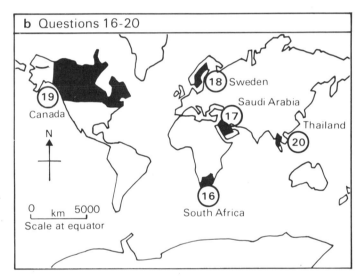

b Questions 16-20

0   km   5000
Scale at equator

World iron ore production

This is a summary of iron ore production in 1985. Draw a bar graph to show these figures. Use parts of the countries' flags (on the right) to decorate the bars of your graph if you wish.

Rank	Production (millions of tonnes)	Chief producer
1	150	USSR
2	75	Australia
3	60	Brazil
4	50	USA
5	45	China
6	40	Canada
7	30	India
8	25	South Africa
9	20	Sweden
10	20	Liberia

Write a paragraph below the graph to say whether iron ore is a finite or an infinite resource, and why.

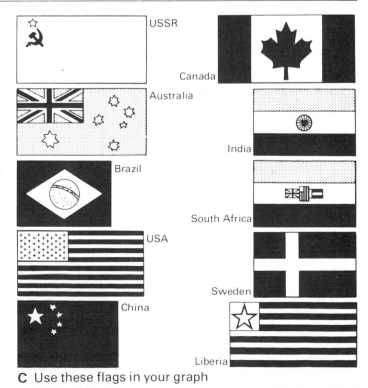

C Use these flags in your graph

Unit 5
Recap

Some places have become very important because of the goods made there. Special conditions are needed if industry is to be successful and make a profit in an area. Industries often attract other industries until a whole manufacturing region is built up.

Study sketch **A**.
Write the correct label for each of numbers **1-5**.
Choose from the following list of labels:

> power . . . goods . . . workers . . . raw materials . . . transport

A Things needed for industry

Write the correct word or words from the list for each of questions **6-10**:

> aerospace . . . profit . . . investment . . . computer . . . manufacturing region

   6 An area where there are many factories.
   7 The money made by selling goods.
   8 Spending money on new factories and machinery.
   9 The industry which makes aircraft and spacecraft.
   10 A modern electronic method of storing and passing on information.

Give the name of:
   11 The busiest waterway in Europe.
   12 The biggest manufacturing region in West Germany.
   13 A railway route through central Asia linking industrial areas.
   14 An island country in Asia which has a lot of industry.
   15 The Florida rocket launch base for NASA.

(Make your choice from the places studied in Unit 5.)

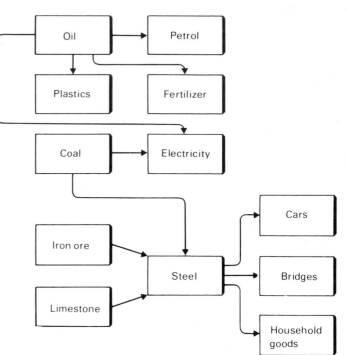

B Flow diagram

Study flow diagram **B**, then complete the following sentences:
   16 Oil is used in cars as power in the form of . . . . . .
   17 Some fertilizers are made from . . . . . .
   18 Car bodies are made from . . . . . .
   19 Iron ore, coal and . . . . . . are needed to make steel.
   20 Oil or . . . . . . can be used to make electricity.

Developing a new area
Sketch **C** shows Landsat taking a photograph of a remote area where there are resources. Geologists have found oil underground.

Write a report for the government to say how the area should be developed. Say what industries you would build, the transport you would use, and what else would be needed. Maps and diagrams can be drawn to illustrate your plan.

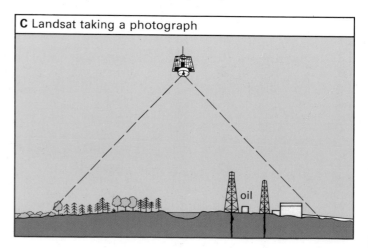

C Landsat taking a photograph

There are many different types of town and city. Each has been built for a reason. Ports are needed to bring goods to and from a country. A capital city is needed by the government.

Some places are more suitable for town growth than others. Some towns grow so big that they merge with nearby towns.

Study sketch A. For questions 1-5, pair up the letters on the sketch with the following terms:

shops and offices
factories and warehouses
suburbs
docks and quayside
truck farms (market gardens)

A Places in a city

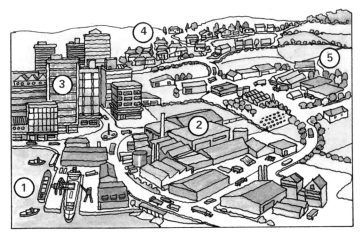

Write out each of the sentences which the characters in cartoon B are saying. Replace the words in brackets with the correct geographical term from the list below:

location . . . densely populated . . . functions . . . site . . . accessible

Which city would you be in if you were looking at each of the following?
11 Red Square
12 Burrard Inlet
13 the barriadas
14 the Kremlin
15 the main US government headquarters (the White House)

Find out the name of each of the cities described in Chart C below. Use an Atlas to help you.

Chart C			
	Country	Sea, gulf or ocean	River
16	China	Pacific	Yangtze-kiang
17	West Germany	North Sea	Elbe
18	Scotland	Atlantic	Clyde
19	USA	Gulf of Mexico	Mississippi
20	Egypt	Mediterranean	Nile

City charades

'City charades' is a team game you could play. The aim is to show the name of a city by acting, or by drawing pictures on a blackboard. The examples in sketch D show how this can be done.

D

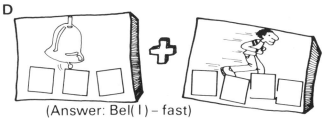

(Answer: Bel(I) – fast)

The game is best played in groups of four or five. Each team thinks of a city, then acts it out in turn. One point goes to the team which guesses the name correctly, and one point to the team which has acted the name.

Select one Atlas map which the whole class can look at: for example, only British cities.

B

6 I'm going to build a town on this (patch of ground).

7 That looks like a good (position) for a town.

8 This place is (full of people living in a small area).

9 That place looks very (easy to get to).

10 What are the main (purposes) of this town?

Unit 7
Recap

Leisure time is very important to us. Making the best use of leisure time depends on having the right kind of leisure provision. This can be a small local park, or a luxury hotel in Greece. Providing for leisure has become an important business in many areas.

Sort out the right ending for each of the following:
1 Beauty . . . . . .   playground
2 Seaside . . . . .   spot
3 Golf . . . .   resort
4 Camp . . . .   course
5 Adventure . . .   site

The luggage labels have been torn at the travel agent. Try to put them together again so that each suitcase goes to the right place.
6 [Dart] [rado]
7 [Tor] [fu]
8 [Dev] [moor]
9 [Colo] [quay]
10 [Cor] [on]

Here are some types of recreation area mentioned in Unit 7:

> a National Park . . . a British seaside resort . . .
> a Greek island . . . a leisure centre . . . a local park

Which would you go to for each of the following?
11 to look at mountain scenery
12 to play on the swings near home
13 to get a suntan and swim in a warm sea
14 to try out new sports and watch others
15 look at shops and enjoy yourself in amusement arcades

Photograph **A**, above, shows Paignton beach near Torquay in Devon.

Pick the best word or words out of the brackets to complete the following sentences:
16 The type of beach is mainly (shingle . . . sand . . . mud . . . boulders).
17 The weather on the day of the photograph was (rainy . . . freezing . . . sunny . . . stormy).
18 The beach is best described as (full of . . . having plenty of . . . empty of . . . having not many) people.
19 Most of the people in the beach are (toddlers . . . middle-aged . . . teenagers . . . pensioners).
20 Most people on the beach are (building sandcastles . . . picnicking . . . playing football . . . sunbathing).

A winter holiday
Photograph **B** (below) shows part of the huge Dillon State Park in Ohio, USA. People can rent the cabins for a week or weekend.

a Make an outline sketch of the photograph.
b Label in the following features:
  Dillon Lake . . . forest . . . hills . . . cabin . . . snow cover.
c Write a paragraph to describe the kind of things you could do during a weekend in winter staying at this cabin.

Earth, water and wind can all move. When they move unpredictably they can cause people great trouble and sorrow. Every year sudden earthquakes, floods and cyclones devastate places. At other times the problem is lack of water.

Give the name of each of the places below from this list:

> Himalayas . . . Ganges . . . Aswan . . .
> California . . . Sahel

1 The giant dam on the River Nile in Egypt.
2 The vast area of Africa often affected by drought.
3 The fold mountain chain squashed up between India and China.
4 The state of the Western USA which is likely to suffer from an earthquake in the future.
5 The Asian river with a densely populated delta which often suffers appalling sea flooding.

Write a word or phrase to answer the following questions. Choose from this list:

> malnutrition . . . Yellow River . . . eye . . .
> Bangladesh . . . tremor

6 What is the meaning of the name Hwang-Ho?
7 Which kind of earthquake causes gentle slow movements of the earth's surface?
8 Name the country devastated by the 1970 Bay of Bengal cyclone.
9 What condition is caused by lack of enough food of the right type?
10 What is the name for the calm centre of a cyclone?

Choose the word, phrase or number from the brackets which completes each sentence best.
11 Cyclones start over (warm land, warm sea, cold sea, cold land).
12 The worst floods are those which are (on land, predictable, profitable, unpredictable).
13 Earthquakes happen (between 5 km and 1000 km, less than 1 km, over 1000 km) below the earth's surface.
14 The areas most prone to droughts are on the edge of (deserts, ice caps, forests, cities).
15 The Hwang-Ho floods in August 1931 drowned (0.7, 1, 2.7, 3.7) million people.

Which terms best fit these descriptions?
16 Vibrations in the earth caused when rocks deep in the earth's crust move.
17 Widespread lack of food, often caused by a prolonged drought.

18 A swirling tropical storm, over 200 km in width.
19 A flow of water over land which is usually dry.
20 An event which brings great trouble and hardship.

Earthquakes
a Make a trace of the world population map **A**.
b Lay it over the map in Unit 8.4 showing where earthquakes will happen (page 66, map **C**).
c Draw rings round the areas on the trace where a lot of people are at risk from earthquakes. Label some of the countries involved. Stick the trace into your exercise book.
d Say why you think people still live in these areas.

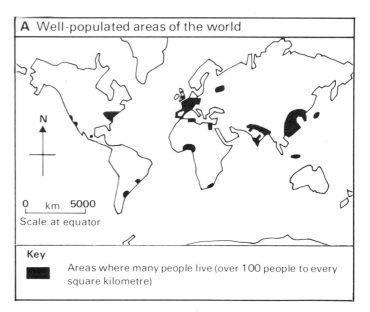

A Well-populated areas of the world

N

0   km   5000
Scale at equator

Key
Areas where many people live (over 100 people to every square kilometre)

Unit 9
Recap

The earth's environment is being changed all the time. People are able to cause great damage to the air, soil, and water. Solving problems for today can lead to more serious problems in the future.

Complete each of the following sentences with the most suitable term:
1 Grassland can turn to   eyesore.
2 Heavy rain on bare slopes makes   desert.
3 Polluted water runs short of   gullies.
4 Ruined buildings are an   restored.
5 Waste tips can be   oxygen.

Write a word or phrase which best fits each of these descriptions:
6 Too much grazing by animals.
7 A heap of useless material from a factory.
8 Huge ships used to carry oil.
9 Land which has been used, then left useless.
10 A large patch of spilt oil floating on water.

'Captain Troubleshooter' goes wherever the earth's environment is in danger. He has been very busy in recent years. Name the places where he has been, using the following clues:
11 A French peninsula to clean up oil.
12 A valley in Wales to restore the land.
13 A very polluted sea between Europe and Africa.
14 A part of Africa being turned to desert.
15 A continent where trees are being cut down too quickly.

The words below are all spelt **wrongly**. Write each one out correctly:
16 **errosion** (to wear away)
17 **vegatation** (plants)
18 **pollusion** (harmful waste)
19 **extint** (died out)
20 **hector** (a measure of area)

The pollution game
Map **A** shows where an oil tanker has hit a reef. Oil is spilling out into the sea. The wind blows it in all directions.

The aim of the game is to see how much damage would be caused. You can play on your own, or against a friend.

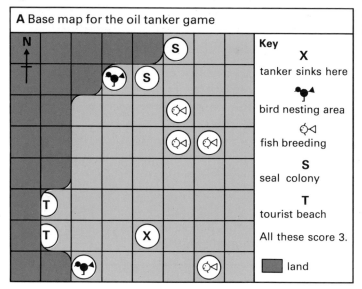

A Base map for the oil tanker game

Wind direction diagram

This is what you do:
a Make a copy of the base map **A**.
b Start the game by selecting any place on the wind direction circle diagram. This tells you in which direction to move the oil slick. (When playing with a friend, you should not both start at the same place on the diagram.)
   Draw a line from the tanker to the centre of the next square in the wind direction you have chosen.
c Now count four moves clockwise round the circle and continue your oil slick line in the direction shown.
d If you hit the edge of the map, or the coastline, count round another four until you can move again.
e After twenty moves, score the amount of damage which has been caused by looking at the chart below:
       1 point for every square you have passed through
       2 points for every square with coastline
       3 points for areas of special importance (shown in the key)
   You only score for the **first** time you pass through a square. The winner is the player with the **lowest** score.

The world is changing rapidly. Many of the changes are for the better. In the past, making a place richer has often meant making it dirtier. Nowadays places can become richer and better to live in than they were before.

Look at sketch **A**. It shows workers from the following list, numbered **1** to **5**. Which is which?

Politician, engineer, surveyor, town planner, forestry expert.

A People who change things

Give the names of the following places from map **B**:

6 The giant port built on a polder near Rotterdam.
7 The capital city of Saudi Arabia.
8 An important river with a large delta in Southern Holland.
9 A volcano which erupted in 1980 near Portland, Oregon, USA.
10 The huge river into which the River Jari flows.

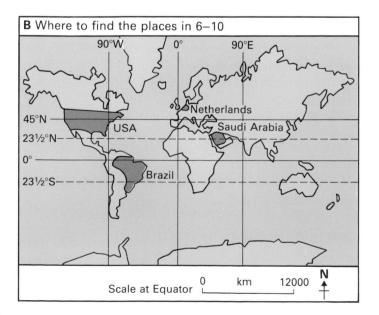

B Where to find the places in 6–10

Give the right word or phrase for these descriptions:

11 Something made at the River Jari.
12 The surroundings of a place.
13 The raw material for the petrochemical industry.
14 Money put into a business on the chance it will make a profit.
15 New farmland created by draining part of the sea.

Use an Atlas to name the countries nearest to these latitudes and longitudes:

16 Tropic of Cancer, 40°E.
17 Equator, 60°W.
18 45°N, 120°W.
19 52°N, 5°E.
20 52°N, 0°.

Improving the quality of life

Look at sketch **C**. Draw it again simply. Label ten changes which could be made to improve the quality of life for people who live here.

C A city scene in the UK

Glossary

Accessible:	Easy to get to/Unit 6.1
Aerial photograph:	A photograph taken from an aircraft/Unit 4.4
Aerospace industry:	Making things which fly in air or in space/Unit 5.4
Ancestors:	Your parents, their parents and so on back into time/Unit 1.1
Architect:	A person paid to design houses/Unit 10.2
Arid:	Dry/Unit 10.2
Avalanche:	A huge mass of snow, ice and rock crashing down a mountain/Unit 8.1
Barren (place):	A place where few plants or animals are able to live/Unit 9.1
Barriadas:	Slums on the edges of South American cities/Unit 6.3
Boma:	Camp for young men of nomadic tribes in East Africa/Unit 2.2
Bore hole:	A hole drilled into the ground to find out what is there/Unit 1.4
Capital city:	The city from which a country is governed/Unit 6.1
Cereal farming:	Farming for grain crops like wheat or maize/Unit 2.1
Coastal plain:	Flat lowland near the sea/Unit 8.1
Coal mine:	Place where coal is brought from underground to the surface/Unit 5.1
Colony:	A country which is governed by some other country/Unit 2.4
Commodity:	Anything which is traded in its raw state/Unit 3.1
Conservation:	Preserving natural resources and the environment/Unit 4.3
Conservationist:	A person who is interested in conservation/Unit 4.3
Corn Belt:	The great lowland plains in East-Central USA/Unit 3.3
Countryside:	Fields, grassland, forest or marsh away from a town or city/Unit 7.2
Crude oil:	Oil from underground in its raw state/Unit 4.2
Currency:	Money/Unit 4.1
(Tropical) cyclone:	A huge swirling tropical sea storm (also called a hurricane or a typhoon)/Unit 8.2
Deficit:	Not enough of something/Unit 3.1
Densely populated:	Lots of people living in an area/Unit 6.4
Derelict:	Once used but now in a ruined state/Unit 9.4
Deserts:	Places where there is little rain and so not enough water for people, animals or crops/Unit 1.4
Disaster:	An event which causes great distress to people/Unit 8.1
Docks:	Quaysides where ships are loaded, unloaded and repaired/Unit 6.2
Drip (irrigation):	Allowing water to drip onto soil near each crop plant/Unit 3.4
Drought:	A long spell without rain/Unit 8.3
Earthquake:	A sudden huge movement of rocks deep underground/Unit 8.4
Elevator:	A large building for storing grain/Unit 3.1
Environment:	Those things which make up your surroundings/Unit 10.4
Exposure:	Poor health caused by being out in cold weather without enough warm clothes or shelter/Unit 1.2
Extensive (farming):	Farming large areas with few farm workers/Unit 3.1
Factory farming:	Rearing animals in special buildings/Unit 3.2
Famine:	Lack of food over a large area/Unit 8.3
Fertile (soil):	Soil which contains plenty of the things which crops need to grow/Unit 9.1
Fertilizer:	Things which may be added to soil to help crops grow/Unit 3.3
Finite (resource):	A thing which can be used up until there is no more of it left on Earth/Unit 4.3
Flood:	A flow of water over land which is usually dry/Unit 8.1
Flow map:	A map showing how many people or things move from one place to another /Unit 7.3
Fodder crops:	Plants grown to feed animals in winter or dry season when there is little grass/Unit 3.2
Fold mountains:	Parts of the earth which have been crumpled and pushed up to great heights by earth movements/Unit 8.4
Gale:	Very strong wind which can cause great waves/Unit 9.3
Game reserve:	A place specially kept for wild animals to live naturally/Unit 9.2
Geologist:	A person who has made a study of rocks/Unit 4.4
Glacier:	A huge mass of ice creeping down a valley from the mountains/Unit 1.3
Gmelina (tree):	A fast-growing tree grown in timber plantations in tropical areas/Unit 10.1

Goods:	Things which are made to be sold/Unit 5.1
Gully:	A very small but steep-sided valley cut by water/Unit 9.2
Harbour:	A place of shelter for ships/Unit 6.2
Hectare:	An area of land just larger than a soccer pitch/Unit 9.2
Ice cap:	Polar regions where ice covers the ground all through the year/Unit 1.2
Infinite (resource):	A resource which will not run out if it is looked after properly/Unit 4.3
Inlet:	A small arm of the sea between two headlands/Unit 6.2
Intensive cultivation:	Farming a small area of land with hard work but with few machines/Unit 2.3
Irrigation:	Putting water onto a field to help crops grow/Unit 2.4
Iron ore:	The rock from which iron is made/Unit 5.1
'Japan Incorporated':	The idea of all Japanese people pulling together like one team to help the nation/Unit 5.3
Landform:	A natural feature with a recognizable shape, such as a hill/Unit 1.3
Landslide:	Sudden movement of tonnes of rock and soil down a slope/Unit 8.1
Lava:	Molten rock flowing from a volcano/Unit 8.4
Leisure time:	Time when a person is not working or sleeping/Unit 7.1
Livestock:	Animals reared by farmers/Unit 3.3
Long profile:	The cross-section of a river from source to mouth/Unit 8.3
Manufacturing:	Making things/Unit 5.1
Manufacturing regions:	Parts of a country which have more factories than other parts/Unit 5.1
Manyatta:	An East African village for elders, women and children/Unit 2.2
Minerals:	Useful resources which come from rocks/Unit 4.4
Mixed (farming):	Animals and crops on the same farm/Unit 3.3
Monsoon:	The sea winds which bring rainy seasons to tropical areas/Unit 9.2
Mountain range:	A line of mountains (e.g. the Ural Mountains in the USSR)/Unit 1.3
Mountain chain:	Several lines of mountains, each one linked to the next (e.g. the Andes in South America)/Unit 1.3
Natural gas:	Gas found in rocks/Unit 7.2
Nomadic (farming):	A way of rearing animals in dry areas by moving them regularly to fresh pastures/Unit 2.1
Oasis:	A spring (waterhole) by which people can live in a desert. (One oasis, two or more oases)/Unit 1.4
Oilfield:	An area of land or sea where oil wells pump up oil from underground/Unit 4.2
Oil slick:	A patch of oil floating on the sea/Unit 9.3
Overgrazing:	Letting too many cattle, goats or sheep feed from an area of land so that it becomes bare and barren/Unit 9.1
Petrochemical industry:	Making things from crude oil/Unit 4.2
Place:	An area made up of land and people with its own special character/Unit 1.1
Plain:	Area of flat lowland/Unit 6.1
(Industrial) plant:	Buildings, machines and tools used to make things in factories/Unit 4.1
Plantation:	A tropical farm growing one or two special types of crop, usually for sale in other countries/Unit 2.4
Polder:	Land which used to be the sea bed until people dammed back the sea and pumped the sea water out/Unit 10.3
Polar regions:	The frozen wastes north of the Arctic Circle (66½°N) and south of the Antarctic Circle (66½°S)/Unit 1.2
Poles:	The most northerly and southerly places on Earth/Unit 1.2
Political:	To do with how some people govern others/Unit 6.1
Pollution:	Harmful filth on land or in the air or water/Unit 1.2
Port:	A water-side town which has a harbour/Unit 6.1
Production line:	A line of people or machines who each add a little bit of work to something which starts at one end as raw material and ends up at the other ready to sell/Unit 5.3
Profits:	The money left after costs have been taken away from the sale of something/Unit 5.1

Pulp mill:	A factory where timber is crushed and mixed with water and chemicals to become mushy pulp for paper-making/Unit 10.1
Quay:	Stone or concrete waterfront where ships can be tied up and unloaded/Unit 6.2
Rain forests:	The great hardwood jungles of the Amazon, the Congo and South-east Asia/10.1
Ranching:	Rearing sheep or cattle in wide open grasslands/Unit 2.1
Raw material:	A thing which other things can be made out of/Unit 5.1
Reclamation:	Bringing something back into usefulness from ruin or from under the sea/Unit 9.4
Recreation:	Active ways for people to enjoy themselves/Unit 7.1
Refinery:	An industrial building where a raw material is purified/Unit 4.4
Risk capital:	Money put into an idea which might or might not make a profit/Unit 10.1
(Crop) rotation:	Growing a different crop in the same field each year/Unit 3.3
Rugged region:	A place which is steep and rocky and has cold, wet, windy weather/Unit 1.3
Sahel:	The dry grasslands south of the Sahara Desert/Unit 8.3
Satellite:	A spacecraft circling the earth/Unit 5.4
Scenery:	The land as it looks from a viewpoint/Unit 1.1
Scientific methods:	Testing an idea carefully to see if it works, and recording the results/Unit 3.2
Services:	Jobs done by people to help other people/Unit 5.1
Shanty town:	Houses built quickly out of odds and ends, usually by poor people on the edge of cities in poor countries/Unit 6.3
Shifting cultivation:	Farming a forest clearing until the soil is tired, then moving on to clear a new bit of forest/Unit 2.3
Site:	The land on which something is built/Unit 6.2
Soil erosion:	The movement of topsoil, caused by water, wind or people/Unit 9.2
Spring:	A waterhole where underground water comes to the surface/Unit 1.4
Squatters:	People who live in empty houses or on empty land without paying rent to the owners/Unit 6.3
Steel:	A mixture of iron, carbon and sometimes other metals. It is tougher than iron and lasts much longer in use/Unit 5.1
Strike:	A deliberate stoppage of work because of a disagreement between the people who are paid to work and the people who pay them/Unit 5.3
Subsistence (farming):	Feeding yourself and your family from your own farm's crops and livestock/Unit 2.3
Super-powers:	The two most powerful nations on Earth, the USA and the USSR/Unit 4.1
Supertankers:	Oil tankers carrying more than 100 000 tonnes of oil/Unit 9.3
Surplus:	Too much of something/Unit 3.1
Surveyors:	People who measure an area and make maps of it/Unit 10.2
Survive:	Stay alive/Unit 1.1
Technician:	A person who is skilled at making something work/Unit 4.1
Temperature:	How hot or cold something is/Unit 1.2
Textiles:	Cloth and clothing/Unit 5.2
Trade:	Swapping a thing which you do not want for something which you do want/Unit 6.2
Traditional (farming):	Ways of growing crops and rearing animals which were the same in your father's day and in his father's and so on/Unit 2.1
Tributary:	A small river which flows into a larger one/Unit 1.3
Tropic:	A line of latitude which is 23½° North (Cancer) or 23½° South (Capricorn) of the Equator/Unit 1.4
The tropics:	The land and sea which lies between the Tropic of Cancer in the north and the Tropic of Capricorn in the south/Unit 1.4
Tsunami:	Giant wave caused by an earthquake in an ocean bed/Unit 8.4
Tundra:	Polar lowlands where ice melts in summer and small plants grow in the cold soil/Unit 1.2
Volcano:	Mountain formed by molten rock coming from deep in the earth/Unit 2.2
Wood pulp:	Crushed timber mixed with water and chemicals/Unit 4.3